The Teacher's Guide to Scratch – Advanced

The Teacher's Guide to Scratch – Advanced is a practical guide for educators preparing sophisticated coding lessons and assignments in their K–12 classrooms. The world's largest and most active visual programming platform, Scratch helps today's schools answer the growing call to realize important learning outcomes using coding and computer science. This book illustrates the expert-level potential of Scratch coding, details effective pedagogical strategies and learner collaborations, and offers actionable, accessible troubleshooting tips. Geared toward the advanced user, these four unique coding projects will provide the technical training that teachers need to master Scratch, feeling comfortable and confident in their skills as they unlock the program's full potential for themselves and their students. Clear goals, a comprehensive glossary, and other features ensure the project's enduring relevance as a reference work for computer science education in grade school. Thanks to Scratch's cost-effective open-source license, suitability for blended and project-based learning, notable lack of privacy or security risks, and consistency in format even amid software and interface updates, this will be an enduring practitioner manual and professional development resource for years to come.

Kai Hutchence is CEO and Founder of Massive Corporation Game Studios as well as its subdivision, Massive Learning, which focuses on educational products and services. Through his established coding support partnerships with elementary, middle and high schools, post-secondary institutions, and provincial and national organizations, Kai has taught over 20,000 students to code and over 2,000 educators to code and teach coding.

The Teacher's Guide to Scratch – Advanced

Professional Development for Coding Education

Kai Hutchence

Taylor & Francis Group
NEW YORK AND LONDON

Designed cover image: © Getty Images

First published 2024
by Routledge
605 Third Avenue, New York, NY 10158

and by Routledge
4 Park Square, Milton Park, Abingdon, Oxon, OX14 4RN

Routledge is an imprint of the Taylor & Francis Group, an informa business

© 2024 Kai Hutchence

The right of Kai Hutchence to be identified as author of this work has been asserted in accordance with sections 77 and 78 of the Copyright, Designs and Patents Act 1988.

All rights reserved. No part of this book may be reprinted or reproduced or utilised in any form or by any electronic, mechanical, or other means, now known or hereafter invented, including photocopying and recording, or in any information storage or retrieval system, without permission in writing from the publishers.

Trademark notice: Product or corporate names may be trademarks or registered trademarks, and are used only for identification and explanation without intent to infringe.

ISBN: 978-1-032-49909-3 (hbk)
ISBN: 978-1-032-50565-7 (pbk)
ISBN: 978-1-003-39905-6 (ebk)

DOI: 10.4324/9781003399056

Typeset in Palatino
by Apex CoVantage, LLC

Additional copyediting and developmental editing provided by
Gui de Souza Rocha.

Scratch is a project of the Scratch Foundation, in collaboration with the Lifelong Kindergarten Group at the MIT Media Lab. It is available for free at https://scratch.mit.edu.

The Scratch name, Scratch logo, Scratch Day logo, Scratch Cat, and Gobo are trademarks owned by the Scratch Team and are used for identification and do not constitute or imply ownership or endorsement by the Scratch Foundation or Lifelong Kindergarten Group at the MIT Media Lab.

With thanks to Gui de Souza Rocha and Terry Hoganson for their assistance, encouragement, editing, testing, and feedback.

Contents

Meet the Author   xii
Foreword   xiii
Reader's Key   xv

1 Introduction   1

2 Our Previous Books in the Series   4

3 Scratch's Place in Coding   7

4 Defining Advanced Scratch   9
   Advanced Scratch Attitudes   10
   Advanced Scratch Goals   10

5 Advanced Project 1: Bar Charts and Data Files   12
   What This Project Is   12
   What We're Learning with It   12
   Building It   14
      Step 0: Creating Your New Project   14
      Step 1: Starting with Random Data   14
      Step 2: The Column Object   16
      Step 3: Generating Clones   17
      Step 4: Scaling Position Dynamically   19
      Step 5: List Stepping to Find the Maximum Value   21
      Step 6: Dynamic Vertical Scaling   22
      Step 7: Adding a Legend   23
      Step 8: Import/Export Data Files   27

6 Advanced Project 2: Point-and-Click Adventure   28
   What This Project Is   28
   What We're Learning with It   28
   Building It   30
      Step 0: Create Your New Project   30
      Step 1: Click to Move   30
      Step 2: Movement Cycles   32
      Step 3: Walk Cycle Animation   34
      Step 4: Perspective with Y Position Scaling   35

Step 5: Setup	36
Step 6: Exits and Scene Switching	38
Step 7: Side Switching	45
Step 8: Starting Up and Wandering the World	48
Step 9: School's Out!	50
Step 10: Sasha and Abby	51
Step 11: The Microphone and Radio	52
Step 12: Noor and the Drumkit	55
Step 13: Getting into the School	56
Step 14: Getting the Guitar	58
Step 15: Ready for the Concert	61
Step 16: Playing the Concert	63
Step 17: More about Sound Effects	64
Step 18: The Main Menu	65

7 Advanced Project 3: Platformer Game — 69
What This Project Is — 69
What We're Learning with It — 69
Building It — 71

Step 0: Create Your New Project	71
Step 1: Ground to Stand On	71
Step 2: Our Moving Player	72
Step 3: Collision Testing	73
Step 4: Jumping	74
Step 5: Drawing the Player	75
Step 6: Starting Positions	79
Step 7: Spikes and Damage States	82
Step 8: Player Lives	85
Step 9: Waypoints	89
Step 10: Exit and New Level	91
Step 11: Bouncer Jump Boosting Platforms	93
Step 12: Ladders and Climbing	94
Step 13: Moving Platforms	97
Step 14: Blinking Platforms	101
Step 15: Victory!	104
Step 16: Title	105

8 Advanced Project 4: Scrolling Shooter — 108
What This Project Is — 108
What We're Learning with It — 108
Building It — 110

Step 0: Create Your New Project	110

Step 1: Player Basics 110
Step 2: Fire Lasers! 112
Step 3: Bomber Enemies 114
Step 4: Damage and Destruction 117
Step 5: Player Health and Death 119
Step 6: Enemy Waves 123
Step 7: Scrolling Background 125
Step 8: Missiles 128
Step 9: Fighter Enemies 131
Step 10: Now with More Lasers! 135
Step 11: The Boss Approaches 138
Step 12: Fighting the Boss 140
Step 13: Boss Movement Patterns 142
Step 14: Boss Health and Boss Phases 144
Step 15: Victory! 146
Step 16: Main Menu 151
Step 17: Leader Boards 153
Step 18: Difficulty Levels 158

9 Advanced Check-In 162
Key Skills 162
  Animation Techniques 163
    Endless Scrolling BG 163
    Stepcount Cycles 163
    Y Scaling 164
    Movement State Graphics 165
    Death Animations/Transitions 165
    Pop-Up Displays 165
  Advanced Variable Handling 165
    Limit Systems (Ammo, Timer) 165
    Child, Parent, or External Properties 166
    List Variables 167
    Cloud Variables 167
    Leader Boards 167
  Complex Movement 167
    Jumping Movement 167
    Inertial Movement 168
    Terrain Interactions 168
    Guide Objects 168
    Inherited Movement 169
    Movement Patterns 169
  Collision Handling 169

Collision Masks	169
Collision Testing Movement	172
Clicking and Collisions	172
Progression Systems	172
Player Lives and Restarts	173
Levels/Scenes	173
Two-Way/One-Way Transitions	173
Level Cloning vs. Level Costumes	173
Waypoints	173
Inventory States	175
Key/Gate Systems	175
Play Dynamics	175
Difficulty Scaling	175
Procedural Generation Basics	176
Clone Waves	176
Dynamic Spawning	176
More Advanced Practice	176
Pet Sim	177
Arcade Classics	178
Life Simulation	178
Teaching Advanced Scratch	179
Teaching Comfort	179
Planning and Designing	180
Testing and Reviewing	181
Analysis and Explanation	182

10 Follow-Up: Extending the Projects   184

Commenting	185
Dinosaur Dance Party	186
Fireworks Display	186
Batty Flaps	187
Butterfly Catcher	187
Pen Tool Fun	188
Interactive Story	188
Snowball Fight	189
Big Map Racing	189
Bar Charts and Data Files	190
Point-and-Click Adventure	190
Platformer	191
Scrolling Shooter	191
Extra Challenges	192

11 Troubleshooting Scratch — 193

- Site Issues — 193
- Coding Issues — 196
  - The Wrong Object — 196
    - *General Tips* — 198
  - The Wrong Block — 198
    - *Confused Pair: Left vs. Right* — 199
    - *Confused Pair: Go To vs. Glide To* — 199
    - *Confused Pair: X vs. Y* — 200
    - *Confused Pair: Set vs. Change* — 200
    - *Confused Pair: Say/Think vs. Say/Think For* — 200
    - *Confused Pair: Play Sound vs. Start Sound* — 201
    - *Confused Pair: If vs. If/Else* — 201
    - *Confused Pair: > vs. <* — 201
  - The Wrong Order — 201
    - *Simple Sequences Code Flow* — 202
    - *Control Structures and Code Flow* — 202
    - *Logic Clauses and Chains* — 202
    - *Concurrency and Race Conditions* — 203
  - Other Errors — 204
    - *Layers* — 204
    - *Visibility* — 204
    - *Colour Selection* — 204
    - *Clones vs. Originals* — 205
    - *Wrong Concepts* — 205
- Backup Plans — 205
  - Offline Scratch — 206
  - Pseudo-Coding — 206

12 Final Thoughts — 208

Glossary — 211
Index — 229

Meet the Author

Born and raised in Regina, Saskatchewan, Kai Hutchence was fortunate to be the son of a math professor (and later research scientist) with an interest in computer science. With his father's help, he taught himself BASIC coding and, as a teenager, HTML, Visual Basic, and other languages. Living in a province with, at that time, almost no tech companies, Kai explored careers in politics, non-profits, and restauraterring before moving to Ontario to work in game development.

In Ontario, he co-founded a game studio that went on to garner over four million downloads of its mobile apps. Living amidst a thriving game development community, he made connections with major players in the industry, helped aspiring creators publish their own games, and saw a community transform into an industry. After almost a decade in Ontario, he moved back to Saskatchewan to take those lessons and try to build up the industry in his home province, a lifelong goal.

In Saskatchewan, Kai launched SaskGameDev as an organization to build, nurture, and support game development in the province. Additionally, he launched his own game development company, Massive Corporation Game Studios. To support the long-term growth of game development in the province, he took up a consultancy role to help spread coding education through schools and non-profits.

While dedicating a portion of his time to teaching, Kai has helped establish coding support partnerships with elementary, middle and high schools, post-secondary institutions, and provincial and national organizations. He has helped develop coding and AI instructional material for nationwide use. Through his multiple partnerships and public profile, he has taught workshops, given lectures, provided industry mentorship and vouching, and facilitated internships with post-secondary institutes across Canada. Following up this success and newfound passion for coding education, Kai launched Massive Learning, a subdivision of Massive Corporation, to focus on educational products and services.

He currently lives in Regina, Saskatchewan. He continues to make games through Massive Corporation and provide educational services though Massive Learning. He enjoys cooking, gardening, and hiking, when he's not teaching, writing, or developing.

Foreword
by Shawn Patrick Higgins

When I talk about my journey as an educator and the lessons I've learned from having taught middle school coding for over 14 years, I often start with the story of the workbook given to me during my first year of teaching. Ostensibly for "teens," it was 314 pages, no pictures, no colour, just page after page a dense wall of text. Undaunted and with a level of enthusiasm that only a first-year teacher can muster, I gave it everything I had, month after month, and while I wish I could claim some passion-fuelled breakthrough, there was none. The year ended with a whimper, and most of my eighth-grade students were sure of only one thing: coding wasn't for them. Perhaps worse still, going into my second full year of teaching, I thought maybe I just wasn't capable enough to teach the subject. It's at this point in the story I usually pause for dramatic effect. I look at the teachers in the room one by one, hold up a copy of that first coding workbook, and suddenly throw it to the floor. "If you want your students to love coding, it's not about completing some stodgy workbook or hitting benchmarks! It's about finding out what your students love and helping them learn to make it!" Fade to black. End scene. Yet another day of CS professional development is done. All the teachers ready to go and lead another successful year-long coding class!

If only it were so easy, right?

I wanted to retell this (true) story to set the stage and give proper context and contrast to the book you now hold in your hands. Of course, all classrooms are different, and one strategy will never fit all students, but Kai Hutchence has written the book that I can honestly say I wish I had when I started teaching.

After more than a decade of helping both my students and my fellow educators embrace creative coding, I've read my fair share of Scratch programming books, but what Kai has done with this series is to forefront a spiralling project-based curriculum that is so good as to make Bruner blush. Woven throughout is an impressive focus on professional-level concepts cleverly nested beneath interest-driven outcomes. Each of the four main projects offers distinctly different lessons and appeals to a variety of learner styles (shout out to the BarChartCrew!), while equipping the reader with contextual tools to understand both advanced coding and design concepts.

I know that some educators will buy this book solely for the higher-level projects that they can immediately implement with their advanced students, but I would strongly encourage all CS and Scratch educators to also read through the end, because both the glossary and troubleshooting sections of this book are some of the best general advice about the logic structure of Scratch code that I've ever seen compiled in a single source.

As Kai says, developing a sense of familiarity with Scratch coding is the only way for educators to be able to authentically understand and speak to the value of their subject. And while there is no substitute for the direct experience of doing the projects yourself, the final chapters of this book comprise one of the best references you will find for elevating your conceptual understanding and, in turn, your ability to effectively support your students.

Near the end of the book, Kai says, "While properly teaching technique might be the standard starting place, truly great teaching inspires students to seek out their own path for the rest of their lives." And after almost a decade and a half of teaching coding in middle and high school, if I've gained any wisdom, earned any cosmic karma, or gathered any guanxi, I would use it all up to hold this ideal as high as I can, with everything I've got. While it's sometimes easy to lose sight of it in the daily roil of the day, what we as educators do can not only change lives but also sometimes save them, opening entirely new life paths our students might not have without us. So to all educators out there, thank you. I'm sure you'll experience just as much joy as I did, joining in on this project-based journey that Kai has so thoughtfully curated for us.

Shawn Patrick Higgins is President of the Oregon Computer Science Teachers Association and 6th- to 12th-grade Game-Making and Coding Teacher in Portland, Oregon. He has over a decade of experience working with youth in creative technology and specializes in project-based curricula that focus on digital art, animation, audio, coding, and games as creative pathways to student success. He is the founder of ScratchEd PDX and has been recognized as 2021 Computer Science Teacher Association Equity Fellow, Processing Teaching Fellow, CS for All Teacher Ambassador, and 2023 Fulbright DAST Fellow.

Reader's Key

Sidebars

Data Structures – how variables relate	15
Magic Numbers – don't pull numbers out of hats!	20
Clicking – mouse events and layering	32
Switching – getting complex with controls	41
Collision – the importance of being grounded	75
Screen Refresh – processing without delay	97
Race Condition – mutually assured destruction	123
Smarter AI – analysis-based reactions	144

Style Legend

To help keep the different concepts involved clear for readers, we have adopted the following text stylings to denote particular things relating to Scratch projects:

Style – meaning
Object – a sprite or the stage
•Code Category – one of the colour-coded categories of code blocks
[Code Block] or **(Code Block)** or **<Code Block>** – any one of the many code components in Scratch (the brackets help convey the shape of the code block)
"Variable Name" – a variable added to the project
//*Script Name* – a connected sequence of code blocks (a stack or script) in a project object

1
Introduction

Welcome to Book 3 of *The Teacher's Guide to Scratch* series! This series was developed as your all-in-one guide to becoming proficient with coding in Scratch and bringing it into your classroom practice. We've covered off an introductory course in *Book 1: Beginner*, our first book, and we pushed into the real heart of Scratch coding in *Book 2: Intermediate*. In this book we'll be covering advanced Scratch coding and getting into the deep details of how Scratch can work.

If you're new to coding, you'll want to start with Book 1 to learn the fundamentals before jumping into intermediate or advanced Scratch. Book 1 has a set of four projects written to guide you through working with the Scratch coding platform, including clear instructions on where to find any component you need. It also has a thorough guide to the Scratch editor and website to help you familiarize yourself with its capabilities and how to work with its core features for coding, digital art, and digital sound to create your own interactive media.

In Book 2 we expanded on the fundamentals of beginner Scratch with another four projects that show intermediate-level Scratch coding. They scaled up the size and complexity of projects and taught important computer science (CS) concepts and methods for handling all kinds of common issues and needs. If you're comfortable with coding, you may still find one or both of these books useful to help understand the Scratch platform and its unique quirks and constraints.

In our third book, we're looking at advanced Scratch. Here we'll be pushing Scratch to its limits with four more projects that will help us explore a few

Figure 1.1 The three books in *The Teacher's Guide to Scratch* series.

last untouched components and introduce a number of more professional CS concepts. Once again, we've changed our approach to writing the instructions in this book to gloss over the minutia of finding things or fully describing every block of code in text. This has allowed us to shorten our descriptions so we can create larger and more complex projects without bulking out the book too much and allows us to focus more on the conceptual understanding of methods rather than the fiddly details of negotiating the Scratch editor. You'll notice the projects in this book are significantly larger and more complex than all our previous projects, with the exception of our one smaller Bar Charts and Data Files project, which is written as a template to jump off from on any number of possible data-centred projects. This project is small but complex and obtuse so is recommended for advanced Scratch users.

The four projects in this book are recommended for students grade 8 to 10, or ages 14 to 16. Anyone attempting these projects should have an intermediate or greater level of Scratch experience and understanding and be able to make their own projects without guidance. These projects will include a much larger number of sprites, or objects, as well as events or scripts, so there will be a lot more diversity in the project, and each object will have more complex behaviours and interactions. It can be a lot to keep track of, so we recommend them only for students already confident in Scratch.

In our first project, Bar Charts and Data Files, we'll take a look at the data science side of CS and explore how we can import, export, alter, and

Figure 1.2 The four advanced projects we cover in this book.

present data in Scratch. It's a relatively simple template designed as a starting point for working with data. It will present the basics of getting and working with data that can then be extended and remixed into any number of powerful and interesting uses. In our Point-and-Click Adventure, we show a more advanced form of narrative-driven game with complex state tracking, inventory system, direct character control, and more interactive elements. This gives a powerful tool for bringing stories to life that allow creators to blend author's narrative creation with player's agency in choosing when and how to explore the world and story. Our Platformer Game provides more work with game physics with lots of examples of how to create interesting movement and interaction systems. It also shows some more powerful methods for animation and state control for more professional-looking, and professional-acting, games and projects. Lastly, our Scrolling Shooter takes us into space to pilot a spacecraft flying through an infinite star field fighting endless waves of enemies and confront a final boss! This project helps teach more about AI, including movement patterns and states. It shows some clever tricks for Scratch to work with clones in complex and interesting ways, even having clones make other clones! We even deal with cloud variables and creating a global high score leader board!

    This book will hopefully guide you through the last remaining untouched concepts in Scratch to help you fully master working with the platform. With the advanced techniques and concepts from these last four projects, we'll review all 12 projects from the series with new ideas about how to revisit them with new eyes and try some new things with these older, simpler projects. These ideas for extra features and revamps are a great way to challenge your most advanced students if they need some more work to keep them busy and challenged.

    Hopefully, this book will help guide you through the last few concepts in Scratch and will leave you prepared and ready to try anything. Our goal here is to take Scratch to its limits and introduce concepts and practices that will leave you and your students ready to make the jump to text coding with professional languages and platforms. So with that, let's dig into advanced Scratch!

2

Our Previous Books in the Series

Before we get started with advanced Scratch, let's take a look back at our previous two books and what we've covered so far. Whether you've been following the series or you're just jumping in at the end here, having a clear understanding of how we've broken down our learning process and reflecting on already-developed skills and principles will help us make sure we're on the right footing for this final part of the process.

Book 1: Beginner was the start of the journey. Here we assumed a complete lack of any experience or knowledge of coding or Scratch. We provided a full breakdown of all the components of the Scratch editor to familiarize with the system we'd be using through the series. This helped to explain working with not just code blocks but also the Costumes and Sounds tabs and all their tools and capabilities that would help us make our future projects. We started with four projects that covered off the basics of how to move using centric or coordinate movements, costume changes, simple control structures with repeat and forever loops, as well as simple conditionals with Ifs and If/Else blocks. We worked with variables and sprite/object properties. This helped introduce most of the fundamentals for CS, but in very simple and straightforward examples.

We saw how animation and sound worked, how we can create harmonized movement, and some simple user inputs in our Dino Dance Party. Our Fireworks Display introduced working with the graphic tools in the Costumes tab, as well as introducing randomization and graphical effects. Our Butterfly Catcher game showed how to work with variables for scorekeeping,

DOI: 10.4324/9781003399056-2

Figure 2.1 The four beginner projects covered in Book 1 of the series.

introduced cloning, and gave examples of simple conditionals and state machines. Batty Flaps showed how to model very basic gravity physics, make more process-based player input reactions, and use a menu system. We used these four examples to be friendly and simple introductory projects to get people started with Scratch. Our aim was to provide easy and engaging projects that could help people overcome the fear and confusion of working with code and see it used for musical, art, or game-aligned projects to provide many hooks for different interest groups. By the end we hoped readers would feel comfortable enough with Scratch to be ready to be explorers with intermediate Scratch.

In *Book 2: Intermediate*, we stepped things up in size and complexity. Our instructions became more succinct, with not as much hand-holding the basics so that we could deal with the increase in project complexity and size. We assumed comfort with the Scratch editor and so could focus on the projects more than the process. Our concept of intermediate Scratch was where the user is familiar and comfortable with the concepts and methods of Scratch and can focus on learning coding techniques for how to handle common problems. They can use these techniques to have a tool set with which they can approach and explore ideas. We wanted to get the user to a point where they can start imagining how they would handle building frameworks of projects mentally, without having to stumble their way through creation trying, testing, failing, retrying of beginner Scratch, to having at least some concept of how they would deal with common tasks or problems.

Figure 2.2 The four intermediate projects we covered in Book 2.

Our Pen Tool Fun project showed off the Pen tool extension for Scratch and allowed us to bring geometry to life. It also allowed us to explore nested loops, a very powerful but somewhat-obtuse coding practice, with nice visual examples. We got to deal with more math functions and practice with nesting reporter code blocks to build math formulae. In the Interactive Story project, we learned about handling text input from users and some important techniques for screening or guiding input. We got to explore animation with larger, more complex, multi-object sequences. This helped provide an example of the creative writing or movie-making side of Scratch excellent for use in English or social studies projects. With the Snowball Fight game, we learned more techniques for handling object motion and physics, creating ballistic arcs for snowball throwing, including direction, power, wind, and gravity effects. We dealt with creating turn structures to limit user inputs and allow computer or AI turns in a game. As well, we added more professional packaging for the game with a Start screen and Game Over screen and the required game state handling to power it. We even learned some sneaky tricks for working with clones to have them collide with themselves or clear out at the end of a level. Lastly, the Big Map Racing project showed us ways to work around the size limitations of Scratch so we could explore a custom-made huge racetrack. We learned some interesting control methods for testing progress and for creating timers or sequencing events like the race countdown. We dealt with referencing another object's properties with the sneaky **([Backdrop] of [Stage])** block and made the player control both the car and the map and move in reference to each other. We made game states to control player controls and control general gameplay so we could add in crashes to the game. Lastly, we introduced a way to have different levels or racetracks and select which map to race on through the main menu. The intermediate-level Scratch was an excellent training ground for a diversity of coding techniques that hopefully set the reader on the path to confidence in working in Scratch and gave them the ability to bring to life any of their project ideas.

If our intermediate Scratch was enough to help the reader bring any project to life, what else could there be? What more could advanced Scratch do beyond that? In this book, we'll build on these already-established skills and concepts by taking things toward more industry standards. We're going to get working with data, dealing with project-wide state tracking, deal more with computer-controlled opponents (simple AI or artificial intelligence), and more polish. So now that we know where we've been, let's talk a bit more about why Scratch is important for coding education and what advanced Scratch is in our next chapters . . .

3

Scratch's Place in Coding

As we approach the limits of what Scratch can do, it can help understand where Scratch fits in the larger picture of the world of coding. I wrote this series of books to help those new to the world of coding and chose Scratch as the best tool for getting started, but where does the rest of the journey lie? This book is about the last stage of using Scratch, so we want to contextualize what Scratch is good for and when you should start looking and moving past.

I chose to make this series entirely focused on Scratch for a lot of reasons. In Book 1, I gave a lot of context about why and how Scratch is the best platform for teachers (and students) to get started with coding. I find it makes coding more engaging, visual, tactic, open, and immersive than competing platforms. Thanks to the socially minded non-profit that runs it, it has a great track record of making decisions that are best for kids. We can count on it as a partner that avoids proprietary licensing, is privacy-respecting, and is free and accessible to everyone, everywhere. It avoids the hassles of unnecessary and expensive physical components that break or go missing. Its platform-agnostic, so kids can access it on the phone, tablet, Chromebook, laptop, or desktop, on any operating system through a browser, or they can even download and install it locally so it can be used when reliable Internet isn't available. I don't believe anything comes close to beating Scratch for how perfectly it suits education.

Scratch is my top recommendation for educators, but that doesn't mean it can do everything for everyone. Scratch is built as a kid's platform, to introduce coding. That goal drove a lot of design decisions, where they always put accessibility and education as top priorities. That means Scratch isn't made for other goals. It isn't suited for any commercial development: there's no Scratch store, what you make there isn't compiled to become proprietary software, and it's built to share and share freely. Because it isn't trying to be commercial software, it can be simple and accessible. Of course, this simplicity comes at the cost of depth of features and complexity. Scratch provides a taste of coding, not a deep, full, and robust development platform. It doesn't include a lot of common CS features or concepts, because they are too complex for the target they aimed for. It isn't going to support a lot of data-centred work with limited data structures or code concepts. Beyond the features included, the way it behaves even has some unusual quirks, where the designers put up safeguards for the earliest learners to keep them out of trouble, prevent objects from moving off-screen or becoming too big or small, for example.

With these limitations, anyone who wants to pursue coding will eventually hit a barrier where they need to move past Scratch. It is a wonderful platform, but it is just a training ground, a place to start your coding journey. Scratch does a good job of setting users up for further CS exploration. As users go through beginner, intermediate, and advanced Scratch coding, their familiarity with coding practices and structures grows. They learn about object-oriented and event-oriented programming without even explicitly knowing they are. They learn the concepts of objects, properties, variables, values, parameters, functions, sequences, loops, and conditions. The fundamentals are all there. By practising with Scratch, they get familiar to code structures, terminology, and syntax, so when they do move to a text coding language, it's not so foreign. They'll understand practices like indenting, scriptifying, and colour-coding inherently from their Scratch projects. Even with its simplicity, it's providing a myriad of implicit and explicit lessons that set them up for future success.

This book will help you explore the upper limits of Scratch, dealing with the last few features and code blocks, of project size and scope that I recommend not going much beyond for the sake of your sanity. Beyond this you'll be ready and able to look toward professional languages like Python or JavaScript, the two most popular text languages for education. Advanced students interested in game development specifically may look to C# or C++ to best suit their long-term interests in learning to work with the Unity or Unreal game engines. For pure education, you might look to the P5JS coding website or the Godot game engine as excellent free and lightweight options to explore text coding. Beyond Scratch lies a vast world of professional-level coding, and you can be confident that having learned and mastered advanced Scratch, you'll be well prepared to take the leap to text-based coding.

4

Defining Advanced Scratch

Advanced Scratch is an open-ended stage of use, pushing it to its greatest possible uses. The question is, then, when do we start considering a user to be advanced? No definition can be hard-coded, with so many different features and purposes; the definitions are as diverse as the users. We can think about advanced Scratch more in how a user works with Scratch than if any singular code block or coding pattern is or isn't being used.

To get to the advanced Scratch stage, a student will have worked through beginner Scratch and intermediate Scratch, which we've dedicated the previous two volumes to. By working through those two levels, we would expect a student that is not only comfortable with Scratch but also confident with it. A user that knows they can achieve what they want and likely immediately grasps the way they'll attempt to approach any given problem and solution (not that they'll always be right first time). We'd expect a student about to begin advanced Scratch to have a familiarity with the goals and outcomes of intermediate Scratch. They should be comfortable with using clones and object references, know how to work with the graphics system, use logical operators and conditionals, and use the messaging system for custom events. If not, then more practice with intermediate would be the best place to start. Our previous book, *Book 2: Intermediate*, is dedicated to the task and should provide exactly the kinds of projects for those learning goals.

Advanced Scratch Attitudes

Advanced students are both the easiest and hardest students to handle. On the one hand, their independence and capability make the general management very easy. However, planning lessons, or worse, troubleshooting can be an enormous challenge with the level of complexity and depth of appropriate projects. Independence is really key; the attitude I expect in students is "I know I can build this". Problems will come up, but they should have the comfort and confidence to push through. While mistakes can and will happen and bugs will inevitably come up, you will have a classroom full of students that can and should be willing to jump on bug-testing and problem-solving. When bugs come up, they are vital opportunities to challenge students at this level. You'll want to engage the class with problems (through sharing projects) and allow students or teams to analyze the problem and find solutions. Ideally, you won't have a quick answer mentality here, but an analysis attitude. You want students to become technicians. Can they understand the project? Can they understand the problem? Can they understand the solution? There will be multiple paths, so having teams develop their own responses and share their analysis, solution, and thought processes is a wonderful and valuable way to handle these problems.

The focus at this level is mastery. They've learned the basics, they've explored the possibilities, and now they need to master things. They need to show the depth of their knowledge. They need to polish their projects. They can take their intermediate projects and refine them, smooth over the rough patches, the small details, and touches to make things professional. Add in menus, in-game restarting, customization options, save data, leader boards, and more. In their code they should be able to develop their own techniques for handling problems. They should take challenges and problems they avoided earlier and try to solve them. As the highest level of Scratch programming, if Scratch can do it, they should be able to try. They won't always be able to get there – there are projects that are truly mind-bending even for professionals – but they should be able to see the general concepts even if they got lost in the weeds of custom blocks of enormous and convoluted projects (which generally would be better suited to other coding platforms).

Advanced Scratch Goals

Of course, as the top level of learning in Scratch, there aren't any areas to avoid, but there are specific areas to focus on to make sure students finish rounding out their Scratch coding education and stay challenged to the end. Our goals for working with advanced Scratch students are:

My Blocks. The most obvious is using custom •My Blocks, which can allow very powerful processing options and dynamic program flow.
Exception Handling. Students should also be able to deal more with exception-handling, working to carefully parse and handle data and conditions. This is very useful for large projects and a key professional skill they can start working with.
Commenting. They need to be able to explain what they are doing and why. Advanced projects should have a lot of comments in them so that others can jump into them and understand the intentions and techniques of the programmer. This is a critical professional skill, and of extreme value in education. Just like showing the work in math, commenting code allows us to see the student's thought process and allows others to understand their projects – a key for good teamwork.
Pseudo-Code. Pseudo-code is where rough plans of how code will be made is done, not writing actual code, but just rough notes that give a sense of the structure and plan for a project. This works like commenting but is done before coding, as a framework for development to plan things out. Again, this provides us critical insight on the student's thoughts and is a critical career skill.
Challenges. In general, we'll need to challenge students at this level, give them new ideas, ask for additional features, or twist or combine other projects. A key technique is to find areas where the student hasn't done something and challenge them to it to try to eliminate blind spots or break them out of patterns and habits.

5

Advanced Project 1: Bar Charts and Data Files

What This Project Is

In this project we'll look at Scratch as a data visualization platform. We'll use it to create a bar chart that can dynamically create data, have it code or hand-entered or imported from data files. We'll create methods that will automatically adjust its size and scaling to learn some important concepts around spacing, planning, and calculation as well as how to work with Scratch's list variables, an important step toward more advanced data processing skills in computer science. It's a simple start project for data visualization that should spur on all kinds of alternatives and customizations. This project should take around 45 minutes to create.

What We're Learning with It

One of the most powerful things computers can do for us is process data. This project is a great way to really start thinking about data processing. This might not seem quite as attention-grabbing as our game-based projects, but with some broad-minded thinking, we can explain the importance of data to making anything and everything with computers. Even games require large amounts of data processing to bring their worlds to life, but these skills can also help us build apps, analyze finances, plan logistics, and run

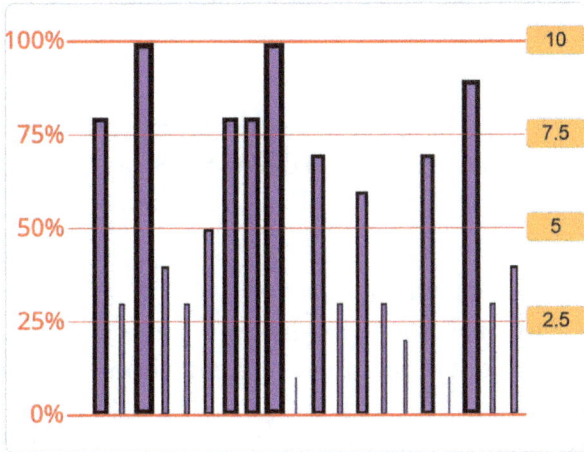

the world. This project will show you how to work with the list variables in Scratch, which are an important tool for complex computing. They are just like other variables, but instead of holding only one value, they contain a list of values. A list variable allows you to hold more data and organize it, with each value within a list given a position in the list, the first value, second value, third value, etc. Computer science has created a number of different ways to store data, called data structures. In Scratch we can learn to master both variables and list variables; in more complex languages, more will be available, but what we learn with lists will be invaluable training for the rest.

> **List Variables.** We'll work with our first list variables to see how they work and try some of the new code blocks for coding with them.
> **Dynamic Scaling.** To adapt to whatever data we need, we'll learn to use some scaling systems for positioning and sizing data representations dynamically.
> **Working with Data Files.** In addition to adding data directly, we'll see how you can import or export data in Scratch using comma-separated value (.csv) files.
> **Minimum Size Handling.** To avoid Scratch protection systems, we'll incorporate a method to allow minimal data values to be represented more accurately.
> **List Stepping.** We'll see how to write code to process data in a list to step through entries to check or modify values.

Building It

Step 0: Creating Your New Project
Make sure you're logged in to Scratch, then click Create to begin a new project! Since we won't be using it, we can delete the Scratch Cat sprite by clicking on the trash bin on that sprite's thumbnail in the Sprite Listing.

Step 1: Starting with Random Data
We're going to start our project by creating some random data. This will allow us to quickly try out things and test them without worrying about what our exact use case is just yet. We'll start in our stage, which will handle the major data processing functions in this project. We'll add a •**[When [R] Key Pressed]** event to allow us to randomize the data on command. Then we'll need to create our list variable.

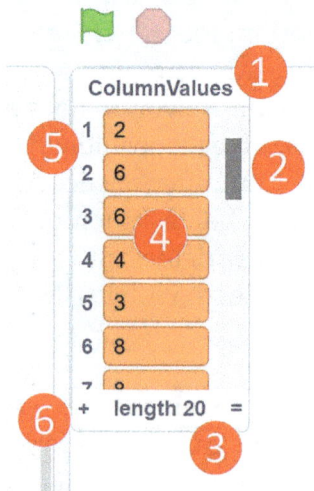

Head to the Variables category and you'll see below the usual variable code blocks a button to *"Make a List"*. Click on this and name your list •*"ColumnValues"*. When you click OK, you'll see a new variable display appear in the Project Window. The display will give the lists ❶ name, then a ❷ scrollable listing of its data (currently empty), and at the ❸ bottom it's length in number of entries. You can reposition or resize this window as you normally would any window. In the Variables category, you'll see you have a checkbox to determine whether this window is visible or not just like a normal variable, and you'll see a lot of new code blocks have appeared to work with the new list variable.

Advanced Project 1: Bar Charts and Data Files ◆ 15

We'll need two blocks for now. We'll start with a ① •[**Delete All Of** [*"ColumnValues"*]] block; this will ensure when we randomize our data, we start by purging the old data from the list and return it to its current "empty" state. We'll also need an ② •[**Add (1) to** [*"ColumnValues"*]] block, but we'll put it in a •[**Repeat (20)**] block. This block adds a new entry to a list and sets its value to the number (or string) provided. Instead of a set number, we'll use the ③ •(**Pick Random (1) to (10)**) block to give us 20 random numbers from 1 to 10 when we run this. Press "R" to try it out, and you should see your list variable now has a length of 20 (the number of entries it holds), and you can scroll up and down to see what value ④ (the number in the orange row) is at what list position ⑤ (the number on the left in the light blue). The add block is the code way to add values, but to enter them by hand, you can click the ⑥ "+" sign on the list variable display to add an entry to the list. You can click any entry to type in a value for it to change them by hand. You can click on an entry and then the "X" on the right to remove an entry, the manual equivalent of the •[**Delete (x) Of** [*"ColumnValues"*]] code block.

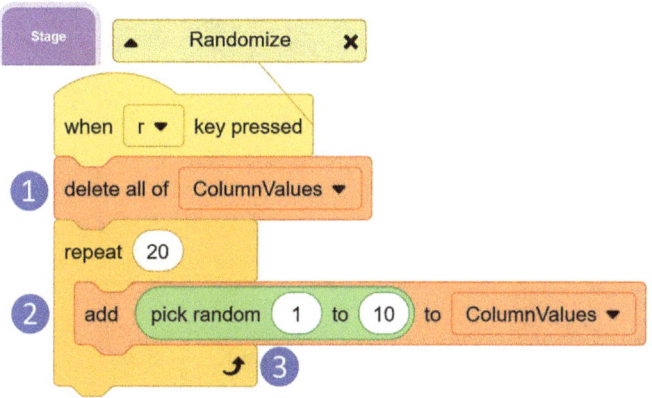

Add (pick random (1) to (10)) to [ColumnValues]

Data Structures

Here we're exploring the list variable in Scratch, as opposed to normal variables. Lists are an example of what's called a data structure in computer science. In Scratch we only use variables, list variables, and cloud variables (see Project 4: Scrolling Shooter for an example), but others exist in other programming languages. Data structures are basically systems to help you work with large amounts of data whose data points relate to each other in different ways. Each data point is like a variable, but depending on the data structure, they'll be able to associate with each other differently, and different code commands will be available to work with them or alter them. In a list, data is arranged sequentially and has a linear integer index, a position within the list. Other data structures include grids, which are like two-dimensional lists with a height and a width; there are also stacks, queues, and maps in most languages. This project is just a tiny taste of working with data, but there's so much more to explore!

Step 2: The Column Object

We'll be using clones to create unique instances to represent each data point graphically. This will require a *column* sprite. You'll want to custom-draw a rectangle, or bar, for this sprite's costume. Try making a rectangle that is 8 × 100 pixels. You can check what size it is in the Costume List to the left. The numbers under the costume name ① are the width and height, respectively, in pixels. Position it so the bottom is aligned to start on the target reticle ② at the centre.

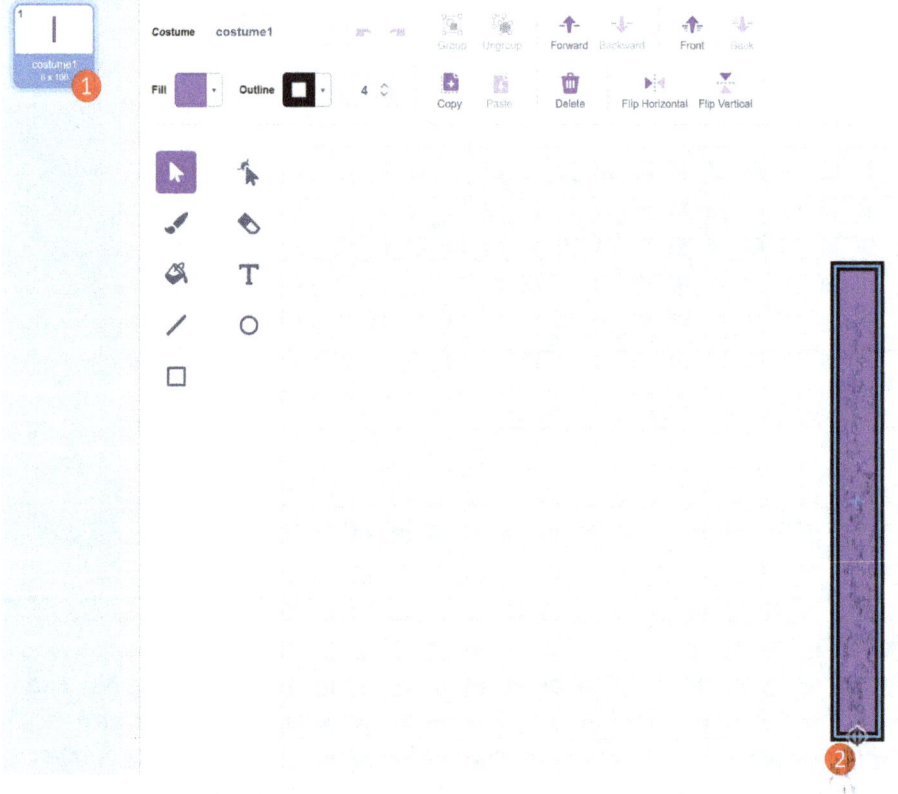

On ▷ we'll have the object ① •[Hide] since we're using clones and don't want the original visible. We'll also use a ② •[When [Space] Key Pressed] to

•**[Delete This Clone]** to purge old clones when we draw a new chart. Then we'll make a ③ •**[When I Start As A Clone]** stack to handle how they should display. This will need to •**[Show]**, since we hid the original, then we'll need a block to position the clone and a sizing block. For now we'll just have default values; our Y positions will always be -150, so they all start from the same place vertically.

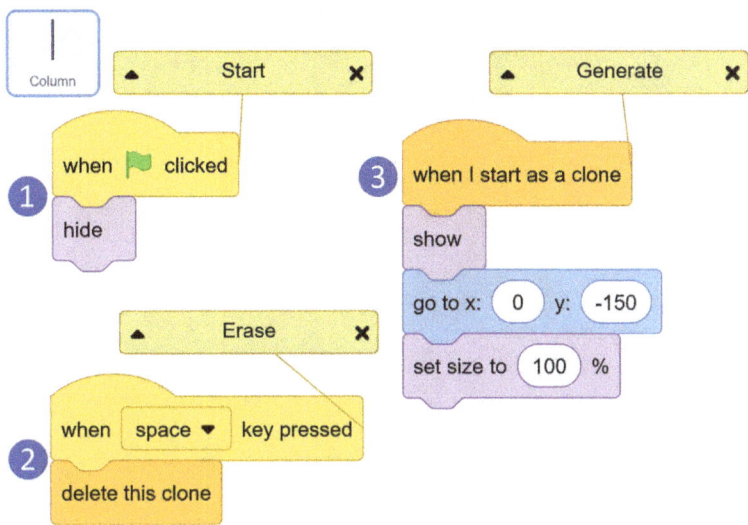

Step 3: Generating Clones

Now that we've got our *column* object, we can start creating some clones. Go to the *stage* and we'll add in a ① •**[When [Space] Key Pressed]** event to handle drawing the chart. We'll start with a delay so that the old clones can delete before we begin creating our new chart. We'll need a new (standard) hidden variable ② •*"Count"*. We'll use this to keep track of where we are in the List index while working through its entries. To make a clone for each entry in our list, we'll use a ③ •**[Repeat]**, but instead of a set number, we'll use a very handy •List Variable code block •**(Length of ["ColumnValues"])**. This block holds the length of the list, or how many entries it has in it. So by using it for our repeat, we'll do something once per entry, making sure our code adapts if the data set changes length. Inside we'll clone a *column* and increase our ④ •*"Count"*, which will be important for allowing us to keep track of our position. A delay will just add a visual effect, allowing the user to see the chart being drawn.

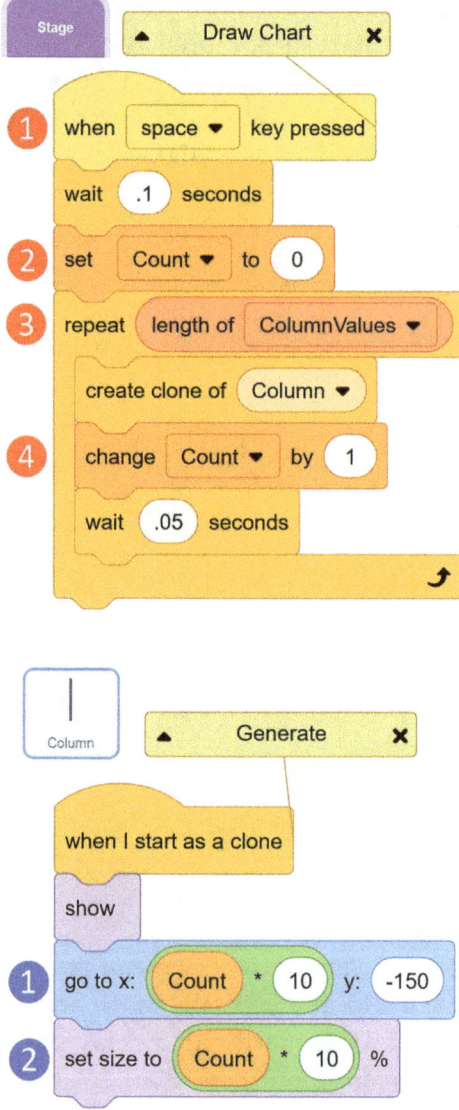

Now, in our *column* object we'll add a couple of tweaks to make things more visible. Our position can now be made dynamic so that our clones don't end up just stacked on top of themselves. Here we can use ① •(•(*"Count"*) * *(10)*) to move ten steps over for each entry. We can also add the ② same to our size to differentiate each entry by size. If you try it out, you'll get a nice, orderly progression of columns – we aren't using our data values yet, just the count or index, so it's just a simple linear progression we're visualizing so far. You'll notice the left-hand side isn't quite the same progression; this is because of Scratch's minimum size protection. Scratch will only let objects get so small; this ensures that users can still see them and click on them. Our object is so thin to begin with it doesn't want to let it shrink down enough. To avoid this protection, add a ❶ wide rectangle to our column costume so it looks like a "+", but we'll make this new rectangle's ❷ fill and ❸ outline both invisible. This makes the shape invisible, but it's still there for size calculations. Try drawing the chart again and you'll see your columns can shrink right down now!

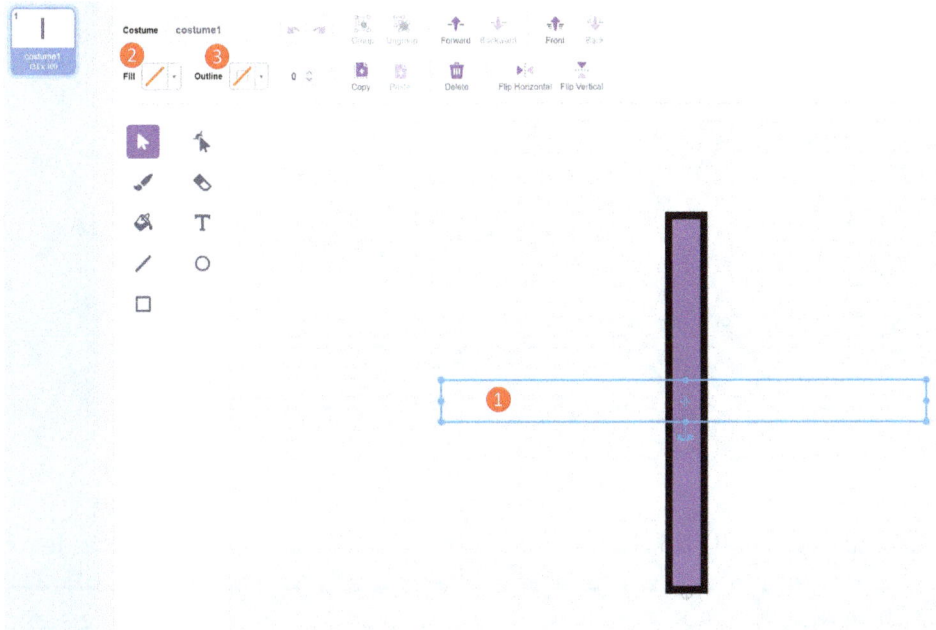

This is a fairly unusual use of clones as data point representatives. If you want a more classic example of working with clones, you can check out the Butterfly Catcher game in Book 1: Beginner, or the Snowball Fight game in Book 2: Intermediate.

Step 4: Scaling Position Dynamically

Our current chart only draws from the centre to the right of the screen; it'd be nice to use more of the space to draw out our chart. Let's add in a couple of standard hidden variables that will help us do that •*"XPos"* to hold the current X position we should create a clone at, and •*"Spacing"* which holds the distance we should put between columns.

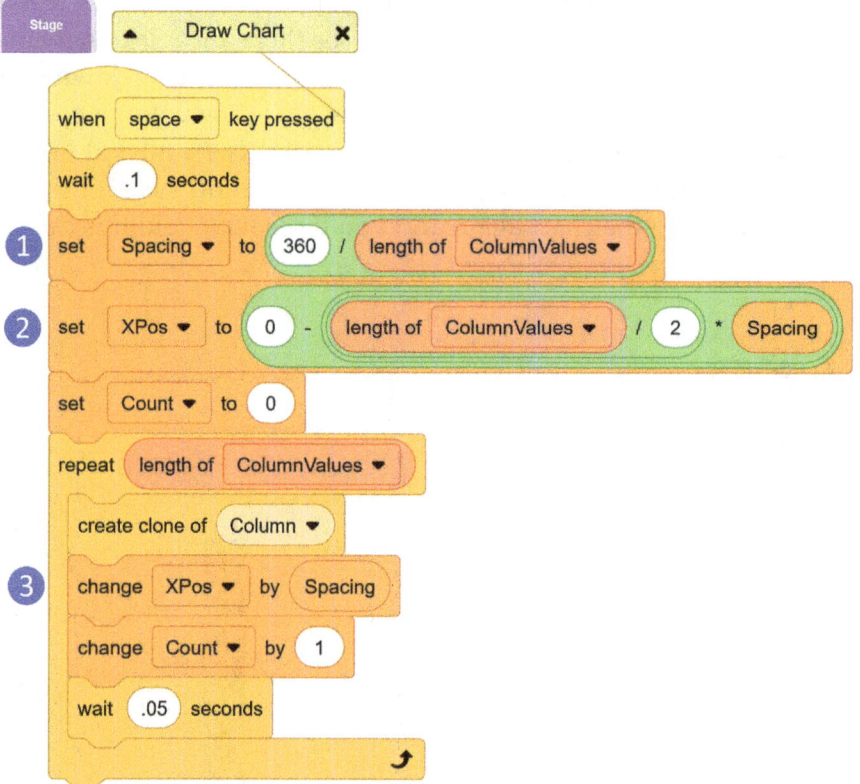

In the *stage* we'll add some code to our *//Draw Chart* stack. After the delay, we'll ❶ •[Set ["*Spacing*"] to •((*360*)/•(length of ["*ColumnValues*"]))]. The 360 represents the total width we want our chart to take up in pixels and divides it among the number of data points (list entries) we need to show. After that, we need to figure out where to start drawing our columns (or spawning clones). We need a ❷ •[Set ["*XPos*"] to •((*0*)-•(•(length of ["*ColumnValues*"])/(*2*)) * •("*Spacing*")))]. This one is a bit of mess, but what we're doing is taking half of the width of our chart and subtracting it from 0 to get to the leftmost position we'll need as our starting point. It might be overkill, but this formula ensures that the measurements will dynamically adjust if anything else changes, which is a handy skill to have and a good practice to get into rather than just typing a hard (or "magic"; see sidebar) number.

> **Magic Numbers**
>
> *"Magic numbers" are a bad practice in computer science. This is a term used to describe when a programmer just types in a hard number that makes things work.*

Everything is built around that one typed-in number, and everything works like magic, but if you have to change that number later, if you've just been typing it out, you'll need to replace all the numbers you typed, instead of using a variable that you can set once and then refer to throughout the code. By using proper variables instead of magic numbers, you don't just protect yourself, but the name of your variable can also be easier to read and give clues about what you're actually doing. Compare reading a "2" verses seeing a descriptive variable name like "number of players". So even though they have a cool name, try to avoid using magic numbers!

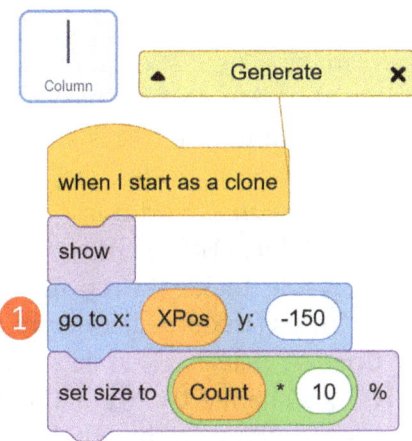

Now that we have our starting •"*XPos*" and our •"*Spacing*", put them to use inside the •[Repeat]. We'll add a ❸ •[Change ["*Xpos*"] by •("*Spacing*")]; this way each entry will move over the •"*Spacing*" amount. With our variables tracking, now we need to edit our *column* to ensure it's using them. In the ❶ •[Go To X: (0) Y:(-150)], add the •"*XPos*" as its X value. Now run your program and the columns should arrange across most of the width of the screen. We aren't using the full 480 width because we'll put the margins to use a little later.

Step 5: List Stepping to Find the Maximum Value
We've got our columns spacing dynamically over the width now, but what about the height of our chart? Data can be wildly different; we could be dealing with % numbers or counts in the millions. How do we know how tall to draw our chart to ensure it can all fit in our Project Window?

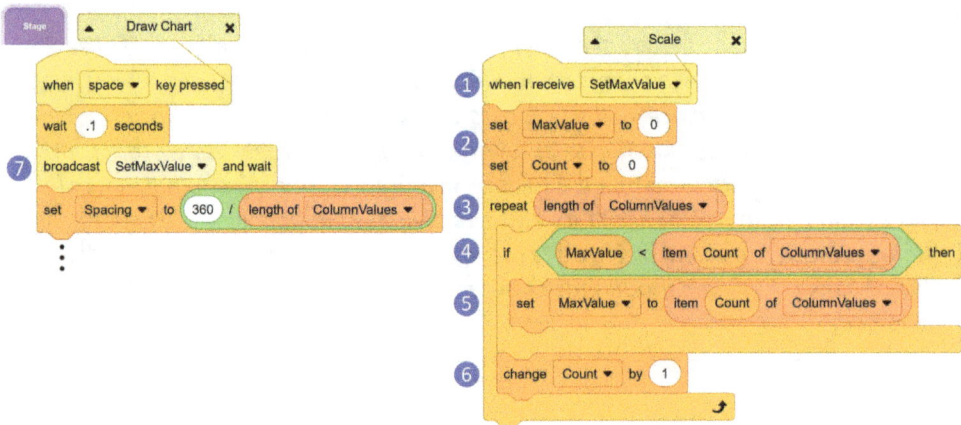

To handle this, we'll make a new stack and event in the *stage*. Add a ① •[When I Receive [SetMaxValue]]. We'll need a new standard hidden variable •*"MaxValue"*. We're going to use this stack to determine the largest number in our data set so we can scale everything to that number. We start ② by zeroing our •*"MaxValue"* and our •*"Count"*. Then we'll use a ③ • [Repeat •(Length of [*"ColumnValues"*])] so we can work through and test each entry in our list variable. Inside the repeat we use an ④ •[If] to test if the current •*"MaxValue"* is less than the list variable at the index position equal to our current •*"Count"*. If it is, ⑤ it sets the •*"MaxValue"* to that new value. After the •[If] we change the ⑥ •*"Count"* by 1 to make sure we step through each index position in the whole list. This is a very common technique, often called stepping through a list or index. You can use the same technique to do all sorts of things, not just check a maximum value; you could test for the minimum, use it to total up the sum value of a list, check how many times a certain value occurs, or all sorts of other uses. Now make sure this new event is called at the start of the //Draw Chart stack using a ⑦ broadcast and wait block. This variation on the broadcast block will call the event, but wait for its event code to be processed before continuing with the code after this code block. This ensures we get the •*"MaxValue"* set before proceeding.

Step 6: Dynamic Vertical Scaling

Now that we have the maximum value, we can put it to use to scale our chart. Let's create a new standard hidden variable: •*"SizeFactor"*. Then at the bottom of the stage's //Scale stack we just created, we'll set our new variable with ① •[Set [*"SizeFactor"*] to •((300)/•(*"MaxValue"*))]. This uses the 300 as the maximum height of our chart. We'll use the same sizing ratio to scale all our other data points in our chart so they are proportional to •*"MaxValue"* being 300 pixels tall.

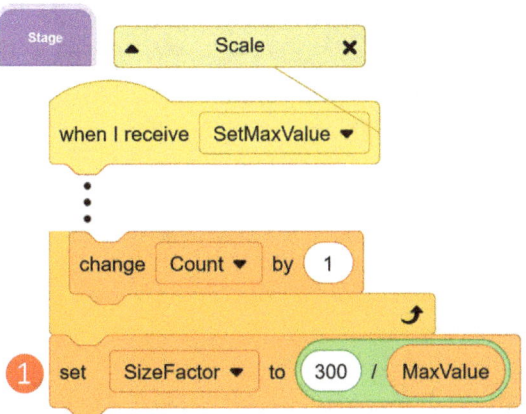

To put it into action, we'll head to the *column* sprite and tweak its code. Delete the •(•(*"Count"*) * (*10*)) from the •[Set Size] block. Instead, we'll use ① •[Set Size To •(•(Item (*"Count"*) of [*"ColumnValues"*]) * •(*"Size-Factor"*))%]. This scales the clone proportionally to the •*"SizeFactor"* ratio. Maximum value entries will be made 300 pixels tall, and everything else will be made proportionally smaller.

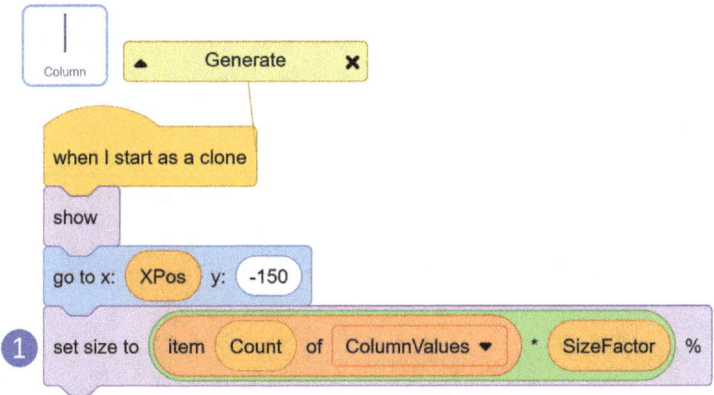

You'll notice that columns are thinner or wider, not just taller or shorter. This is because the size of a sprite in Scratch is rescaled both in width and height proportionally; other programming languages can often scale just the height or width of an object. There are other options for how you draw out the data than just scaling a sprite, so you can challenge yourself, or your students, for alternatives.

Step 7: Adding a Legend
Of course, any good chart needs to have a legend to aid the audience in interpreting it. We're going to add not just a legend but a reactive dynamic one, but

we'll need a few objects and variables to do it. In the *stage*, add a new message broadcast to the bottom of the *//Scale* stack: ① **[Broadcast (Rescale)]**. We'll use this to trigger our legend to rescale. Create four new standard visible variables: •*"100"*, •*"75"*, •*"50"*, and •*"25"*.

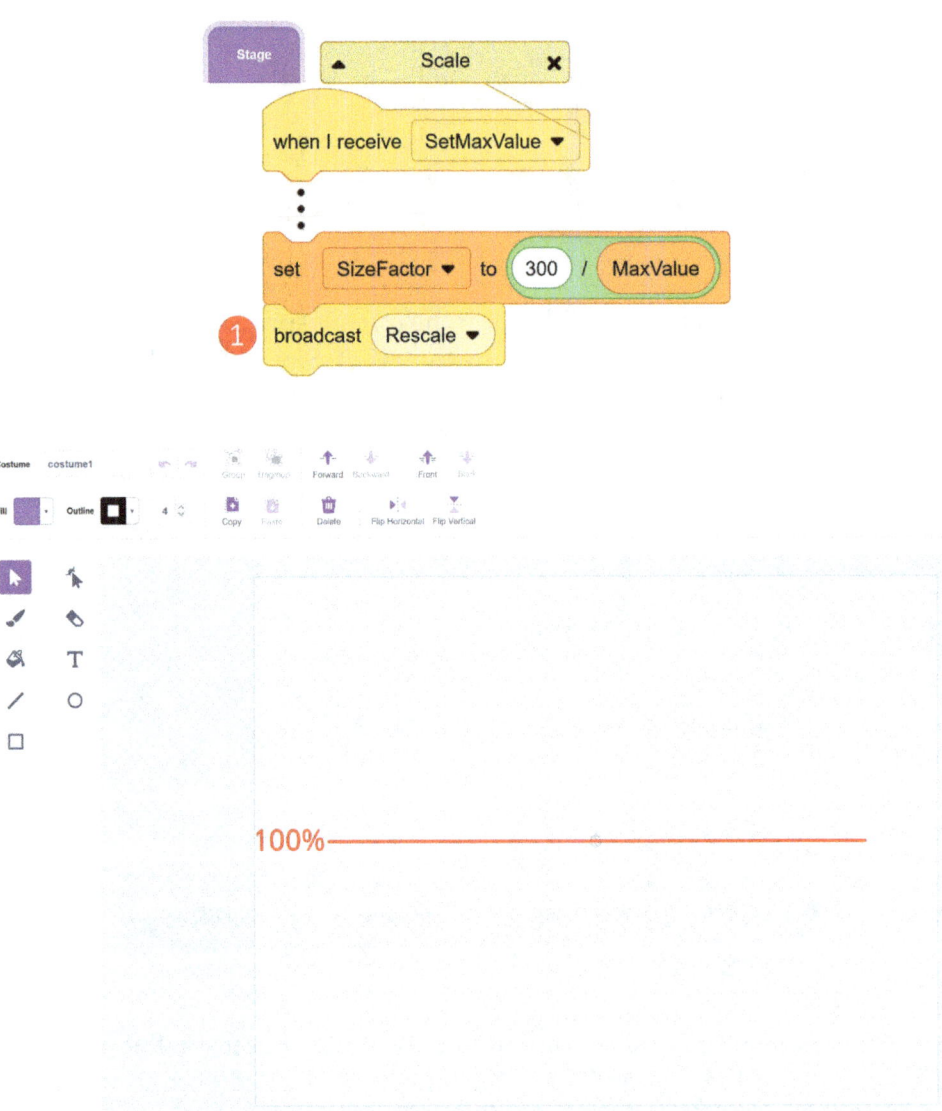

Now we'll create a new sprite with a custom costume. Let's call it *100%*. This object will be our 100% or maximum value line. Its costume should be a thicker line (maybe 4pt) across most of the screen, with "100%" in text at the left-hand side. Make sure the line and text are aligned over the target reticle. The code for it will position it ① at X: 0, Y: 150, will set the ② •*"100"* variable to the •(*"MaxValue"*) and show it. Click the ▷ to position the line. Right-click on the •*"100"* variable display in the Project Window and select "Large

Advanced Project 1: Bar Charts and Data Files ◆ 25

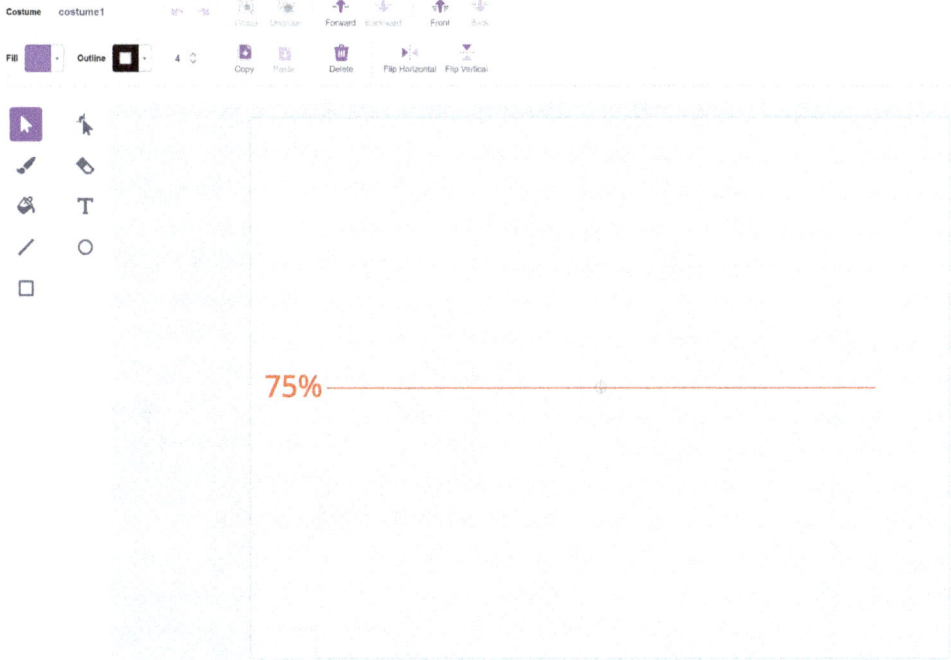

Readout" style and then position it in line with the 100 line but at the right-hand side. We now have an indicator for our chart of the maximum value in both % and real numbers.

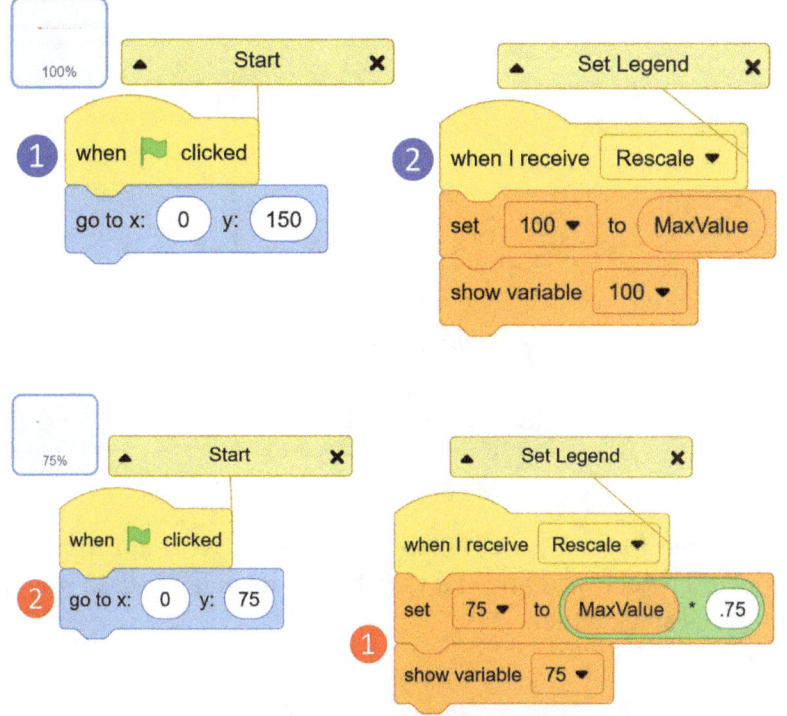

Next, we'll duplicate the *100%* sprite to make a *75%* sprite. In this one, change its costume to read "75%" and make the line only 1pt thick. Adjust its variable readout, and importantly, we want the variable to ❶ •**[Set [75%] to** •(•("*MaxValue*") * (0.75))]. Its position should be ❷ Y: 75.

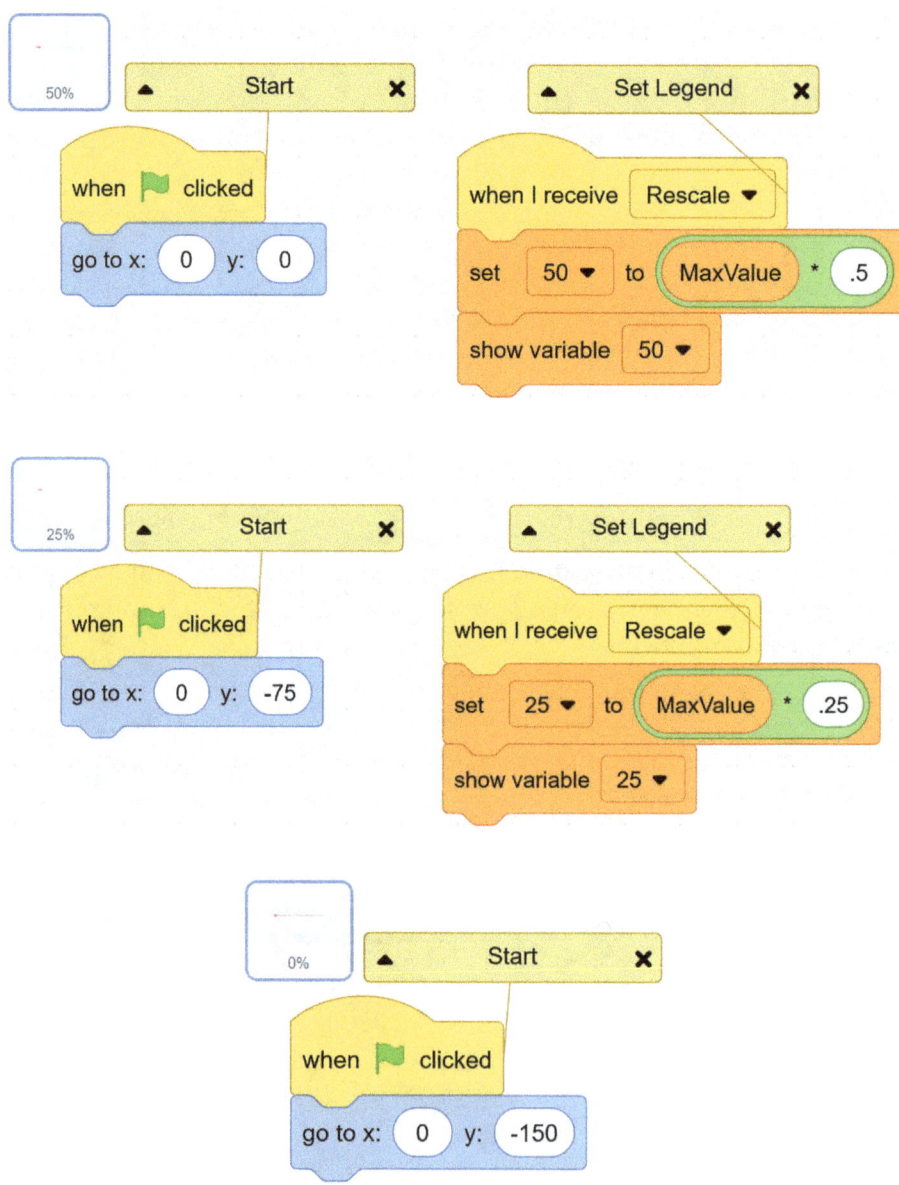

Do the same for a *50%*, *25%*, and *0%* objects with appropriate adjustments. The *0%* object won't need a variable, as its value is 0, so it won't need the *//Set Legend* stack either.

While most folks probably don't immediately think data representation when they think fun coding projects, data visualization is an important field and can be fun and artistic. For some more techniques to work with visualizations, check out the Pen Tool Fun project in Book 2: Intermediate.

Step 8: Import/Export Data Files

There's one last feature of list variables that we want to explore in Scratch for this project. If you make your list variable visible, you can right-click on the list variable display. You'll see you have three options: ❶ Export, Import, and Hide. *Hide* of course simply hides the display. *Import* and *Export* allow you to work with data files. To give yourself a view of what a data file for Scratch is, choose the Export option; this will allow you to download a copy of your current list variable's data. You'll see it's simply a text file with each number on a separate line, as minimal as can be. This might not seem that useful, but for dealing with large amount of data, this simplicity is very handy. You can make your own data files to import. You can use spreadsheets or text editors to make your own files. Save them as either. txt files or as. csv files and they'll be able to be imported into Scratch. It might not seem like much now, but you'll find working with data files is a lot more convenient than using code or manually adding data within Scratch.

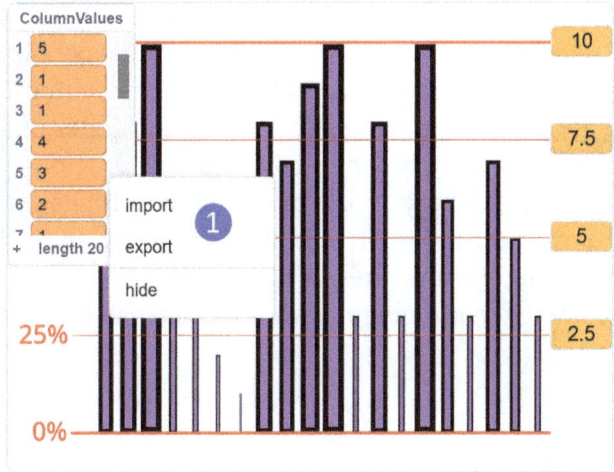

For working copies of this and every project in the book series, visit www.massivelearning.net for direct links to Scratch Projects, and to see our other projects and resources for coding education!

6

Advanced Project 2: Point-and-Click Adventure

What This Project Is

For our first advanced game project, we'll be looking at point-and-click adventures. This genre of game combines the storytelling from our earlier Interactive Story project with a more open-ended flow, with players taking direct control of a character to freely move and interact with people and objects in the environment. With no set linear path but rather needing to adapt to the player moving around and making decisions in any order of their choosing, this is a step up in complexity when compared to previous projects. Here, we'll follow the story of Avery trying to get her bandmates ready for a concert while wandering around town, meeting different characters and dealing with many objects. In around three hours, we'll end up with an adventure story full of interactive components and a framework for tracking progress through the story.

What We're Learning with It

In our first advanced project, we'll be looking at a number of new uses of previous techniques, expanding on their premise to create deeper and carefully controlled processes. By exploring some common industry techniques, we'll build more accurate and synchronized animations. We'll work with more state machines to control not just game scenes but character and story

aspects. Lastly, we'll learn more techniques for creating interactions between objects and narrative or game states. This is the largest, most ambitious project so far.

Stepcount Cycles. A very useful processing system for animations and other timed effects.
Click Collisions. A technique to handle mouse clicks anywhere in the game and more ways to work with our mouse as a controller.
Room/Scene Transitions. A new technique for making room-to-room transitions more action-based and visceral.
Y-Scaling/Perspective. A system for adding some perspective to our games through object scaling by position.
Pop-Up Displays. Two different information displays in-game to provide the player (and designer) with visual feedback about game states.
Inventory States. Tracking of inventory states to help narrative development as well as information for story advancement and displays.
Key/Lock Story States. Track the advancement of the storyline and have the game world's non-player characters (NPCs), scenes, and objects respond dynamically.

Building It

Step 0: Create Your New Project
Make sure you're logged in to Scratch, then click Create to begin a new project! Since we won't be using it, we can delete the Scratch Cat sprite by clicking on the trash bin on that sprite's thumbnail in the Sprite Listing. Then you can add in the backdrops we'll need for the project and rename them in the **Costume** tab to keep them straight:

- **Bedroom3** (renamed Avery's)
- **Woods and Bench** (renamed Park)
- **Soccer**
- **Room 1** (renamed Abby's)
- **School**
- **Theatre 2** (renamed Backstage)
- **Basketball 1** (renamed Basketball)
- **Concert**
- **Spotlight**

Step 1: Click to Move
Add the sprite *Avery* to act as our main character. We'll set Avery's rotation style to left–right, matching the movement style we're going for: mainly a side view of the world with some perspective depth from the bottom of the screen to around the middle of the screen. Go into the **Costumes** tab and move Avery's image so the ① feet are on the centre target, and at the top of the Costumes List, add the ② four costumes from **Avery Walking** by clicking on Choose a Costume in the lower left corner and adjust those costume's positions as well.

Add a new *"GoTo"* event, and under it add a ① *[Glide 1 Second to X: ("GotoX") Y: ("GotoY")]*. These two new hidden variables will be used to create a target coordinate for our movement. To set the movement targets, go to the *stage's* code.

In the *stage* we can start with a ① *[When ▷ Clicked]* and [Switch Backdrop to (Avery's)], so we start in her bedroom. While we're here, add a ② *[Broadcast ["GameStart"]]* after that because later we'll be adding the *"GameStart"* event method to kick everything off later on. To move *Avery*, add a ③ *[When Stage Clicked]* event. Under it, add in a ④ *[Set ["GotoX"] to (Mouse X)]*, a *[Set ["GotoY"] to (Mouse Y)]*, and a *[Broadcast ["GoTo"]]*. Now, test it out. Wherever you click, *Avery* will move until she reaches where you clicked.

Advanced Project 2: Point-and-Click Adventure ◆ 31

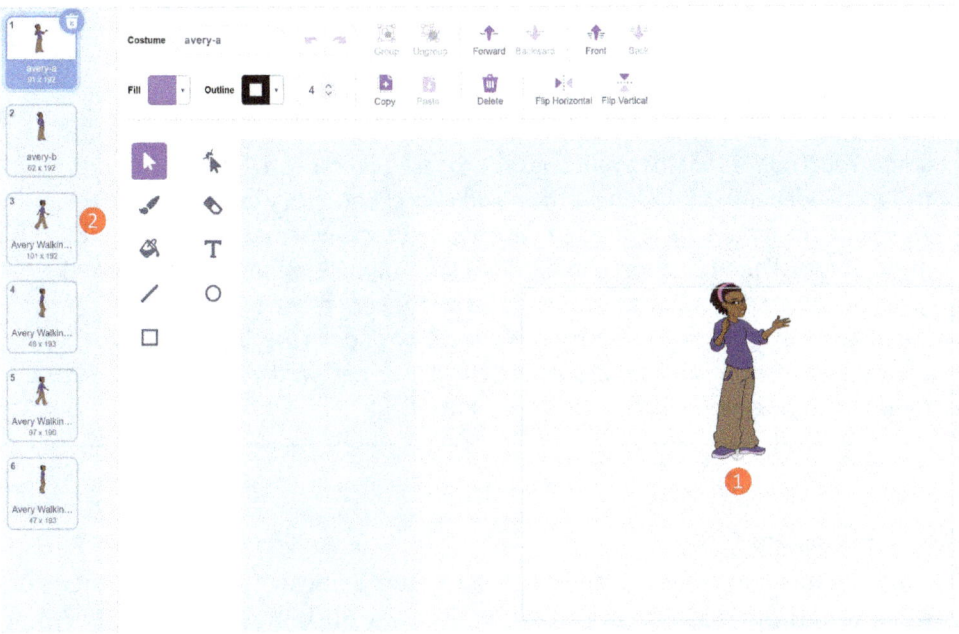

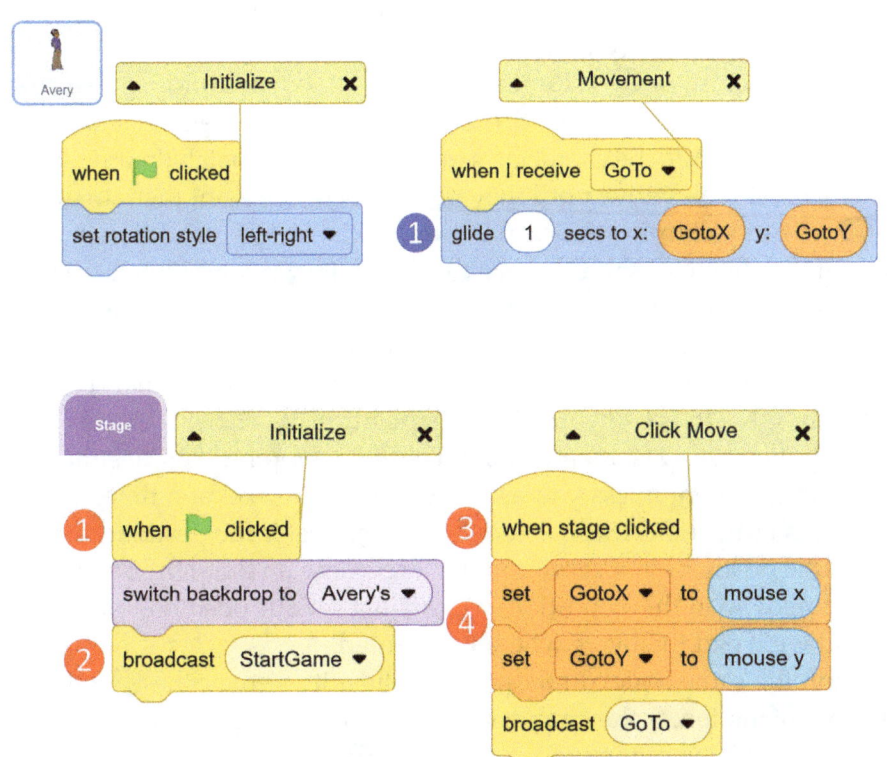

> **Clicking**
>
> *This system uses a few features in Scratch that work with the mouse. We've used When Sprite Clicked a lot before, and its fairly obvious to use, but here we use the stage. You'll notice that the event doesn't trigger if you click on Avery. Sprites are all independent objects that have their own click event; if they are in between the stage and the mouse, they intercept that event instead, whether they have a click event or not. Whatever object is in front will take the click event, not any others. We grab the Mouse X and Mouse Y coordinates on the click event, which are the Scratch coordinates that align with the mouse's current position in the Stage Window. They can be used for exact coordinates, or you can use those numbers to calculate relative motion to create motion effects or other controls.*

Step 2: Movement Cycles

The basic click-to-move system created in Step 1 doesn't support change costume animation (it looks more like, well . . . a glide than a walk). For that, we'll create a step-by-step motion system. Fair warning that it might look big and intimidating at first glance, but it's just that we need four different copies of the same thing to handle each different direction of movement. Go back to *Avery* and remove the **[Glide]** on our **"GoTo"** event.

We'll make use of ① **[Repeat Until <<<(x position) =・("GotoX"))> and ・<(・(y position)=・("GotoY"))>>]** to find our movement goal. We are running a loop while the character's position doesn't match their destination, so this will let them move until they get to where we want them to go.

② **[If <・("GotoX") < ・(x position)> Then]** is a conditional test that checks if the character should move to the left. It evaluates if their destination has a lower value than their current position, since in Scratch the X position increases to the right; a lower X position is to the left. In this condition, we want to face left and move left, but we need another conditional so we don't go too far. ③ **[If ・<(・("GotoX") > ・(x position))> Then]** works as a safety feature, if they step past their destination. Overstepping your destination can happen fairly often unless you are only moving 1 pixel at a time. The destination was to the left. After we move, we then test if the destination is to the right. If the destination that was on the left is now on the right, we must have overstepped! So in this condition, we can just set our position to the destination and we'll end the move where we wanted.

Using the same logic, duplicate and adjust it to cover the ④ other three directions of motion. In the up and down movement, don't bother with a **[Point in Direction (#)]**, since the sprite doesn't have art for up or down movement.

Advanced Project 2: Point-and-Click Adventure ◆ 33

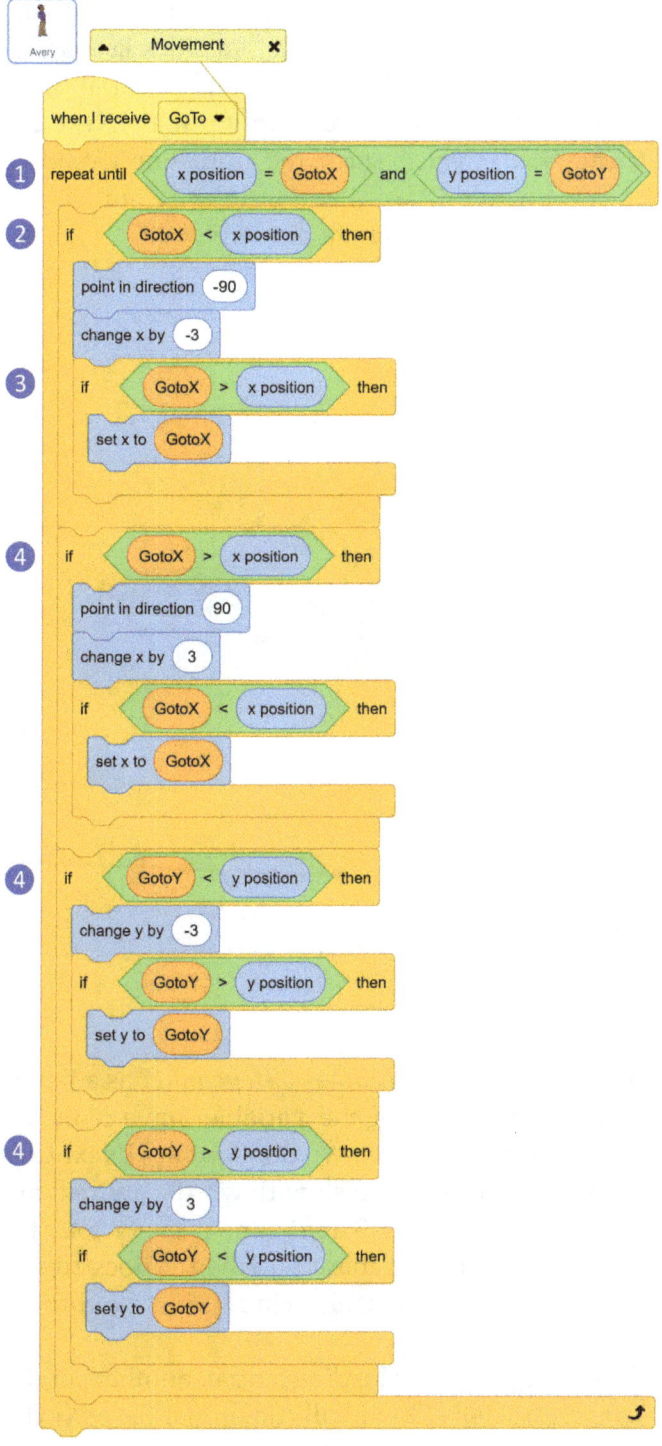

Step 3: Walk Cycle Animation

Now that we've got the movement stepping toward the destination, we'll animate the character. Having it in its own custom block keeps the code cleaner. Start adding a ❶ •[*"WalkCycle"*] my block and call it at the bottom of but inside the •[**Repeat Until**] stack. If we simply put a •[**Next Costume**] block in the ❷ •[**Define** *"WalkCycle"*] stack, the animation would be too fast and include costumes we wouldn't want in a walk cycle animation. Therefore, it demands a little more work to animate properly.

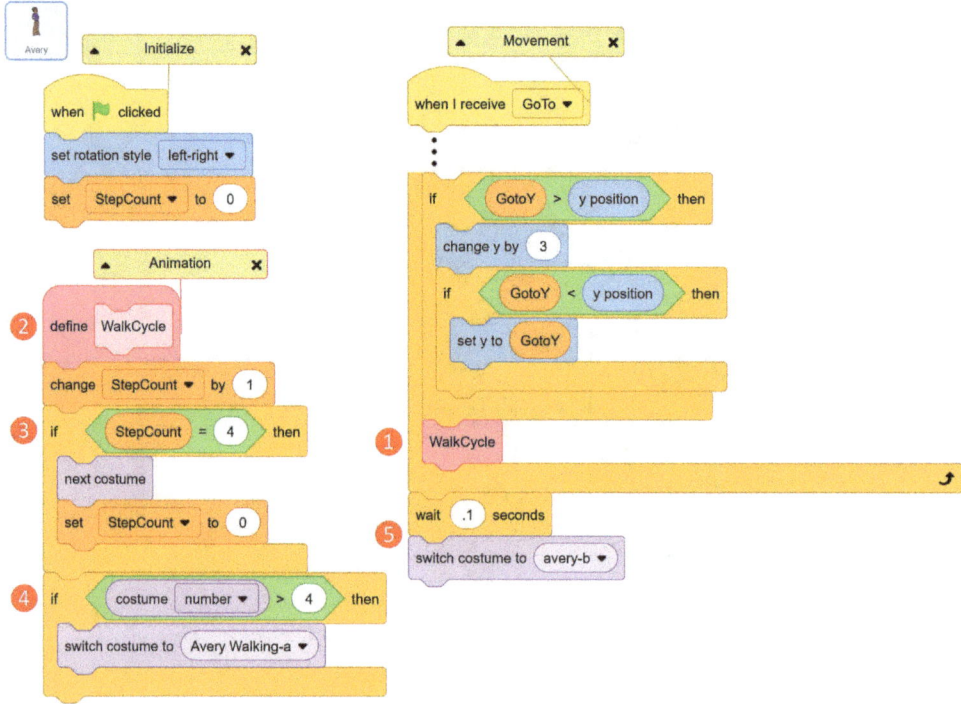

The trick is to use a hidden variable •*"Stepcount"* as a timing mechanism for the animations. After creating the variable, have it increment (change by 1) each •*"WalkCycle"*. However, we don't want •[**Next Costume**] to run every frame, so use •*"Stepcount"* to slow down the animation cycle. Setting an ❸ •[**If** •<•(*"Stepcount"*)=(4)> **Then**] runs it every four frames instead. Next, reset the •*"Stepcount"* variable to 0 when it does run, so it can hit 4 again in another four frames/calls, allowing it to cycle through our animation as needed.

To avoid the wrong **costumes** loading, add in another conditional. ❹ •[**If** •<•(*Costume Number*) > (4)> **Then**] will run only if the •[**Next Costume**] has pushed the costume into the unwanted ones; in that case, it resets the costume to the start of the cycle. Now, *Avery* can walk anywhere, has an appropriately timed animation cycle, and only uses the right costumes. A ⑤ •[**Wait (0.1) Seconds**] and •[**Switch Costume to [*Avery-B*]**] below the •[**Repeat Until**] will even switch *Avery* to standing still when she's done moving.

If you'd like to have a simpler system of animation to start with, you can check out the Dino Dance Party for some tips in Book 1: Beginner.

Step 4: Perspective with Y Position Scaling

Most of our backgrounds are drawn with some amount of perspective. There's a simple trick to have our characters match that perspective. Keep in mind that since we're using a whole bunch of different backgrounds from the library and they're drawn in slightly different styles, your results may vary depending on the backdrop. You'll be able to use this technique to fit other projects more precisely if you have more control over the art, so keep this concept in mind.

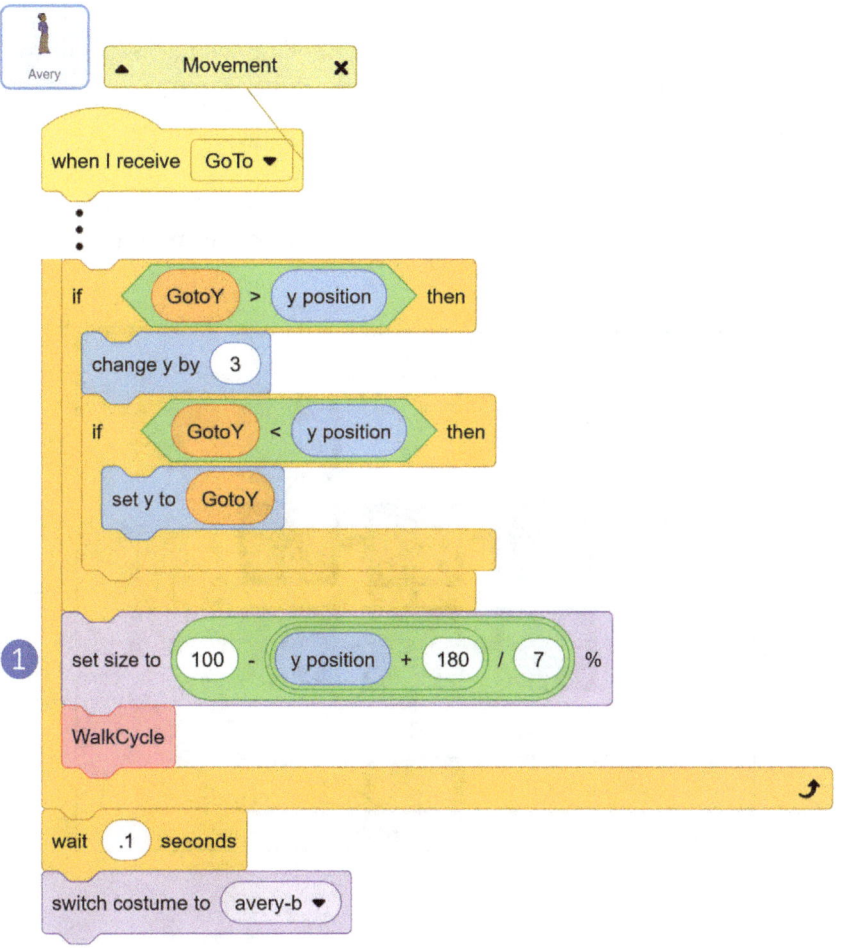

In Avery's code, jump down to the ●[*"WalkCycle"*] call. Just above it, add a ① ●[**Set Size To** ●((*100*) − ●(●(●(**Y Position**) + (*180*))/(*7*)) %]. We're changing size, but determining the right size to match the perspective takes a bit of guesswork. Change the divisor to suit your use, and the other portions of the formula will remain.

The ●((*100*) − ( )) gets a portion of the normal size, assuming 100% is the maximum you want the character to be resized. We'll deduct some amount from that.

The ●(●(**Y Position**) + (*180*)) is how we determine the character's position, but what's the +180 for? Because of the Scratch grid, the very bottom of the screen (the lowest point you can click and move to) is -180. We +180 it to compare any position to the minimum of 0 at the bottom; every pixel up from there is now a positive integer scaling off, having zeroed our scale from -180.

Then, take however many pixels up from the bottom the character is and divide it by 7. With the screen being 360 pixels tall, we'll get at most 360/7 = 52.43. If we minus that from 100, we get 48%; our character will range from size 100% at the very bottom of the screen to size 48% at the very top. Changing the 7 divisor will give you a stronger or weaker scaling to match different perspectives.

Step 5: Setup
In any project, but especially a big one like this, there are tried and true ways of helping remember what you're doing and where you're going.

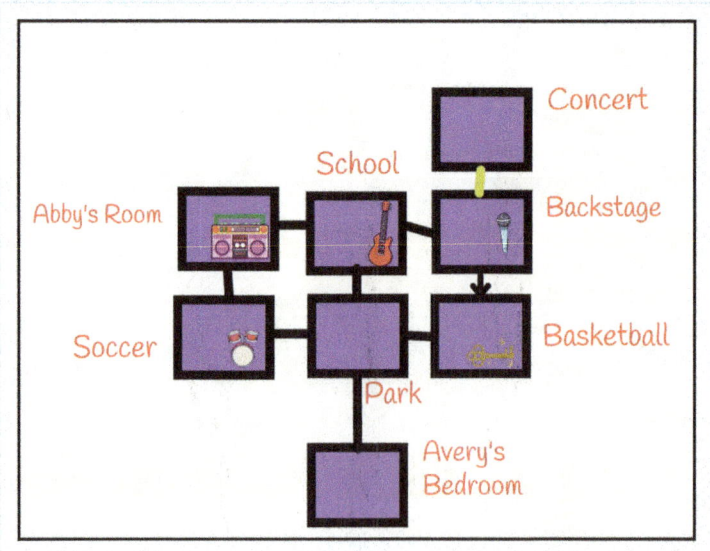

Something very common in game development is using a debug mode, running the game in a way that helps developers keep track of things that you don't see in the finished project. You could, for example, leave a variable visible on-screen to make sure your code is working, by seeing what that variable is doing throughout gameplay. To have a better view of what the project will become, we'll be putting together a map of our game to help us remember our map layout.

Create a blank sprite and follow the example to create a similar **costume**. By copying in a number of other sprites that we'll be using in our project, we've placed them into the scenes they'll be located in, naming each scene to match the **backdrop** we'll use for it. The line between scenes shows how they connect. A yellow line indicates a locked connection, and an arrow indicates a one-way connection. This will work as the *secret map* of what we're building.

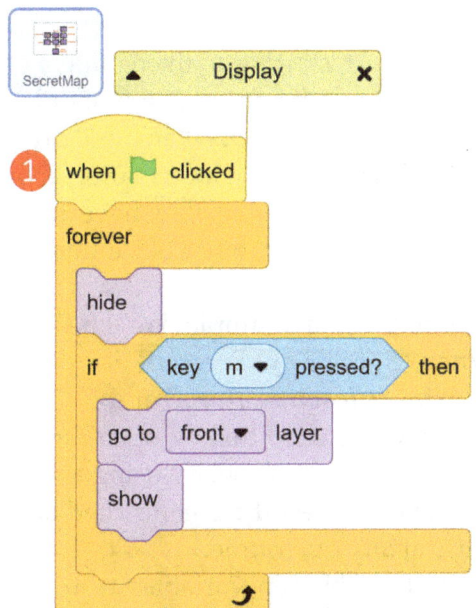

Next, by adding some code to our secret map, we can call it up in-game. In a ❶ •[When ▷ Clicked] and •[Forever] loop, we'll •[Hide] the sprite, but if the "M" key is being pressed, it will display. "M" for map, nice and easy to remember. Now we have this handy reference while we build our game. It's up to you whether to include it in the final project or to delete the sprite before sharing.

Add in an *inventory* object, starting with creating a blank sprite. It'll need a costume with spots for five different items. It should have text to explain what it is, as well as instructions to click when done to exit the inventory. This will be used to give the player access to information about the state of play in an in-game context. They'll see the items they have acquired, though it won't tell them what to do with them and what place they are in the sequence of events involving that object.

Code for the **inventory** is fairly simple. We'll need a •**[When This Sprite Clicked]** and a •**[When [Space Key] Pressed]** events. They'll control its visibility but also call a ❶ •*"CloseInventory"* or ❷ •*"ShowInventory"* event, respectively. We'll put a single condition on opening the inventory to ensure that the player isn't in the concert scene. By using a ❸ •**<Not <condition>>** logic code block, we can easily test if the concerned backdrop isn't currently in use, allowing its use at any other time.

Next, we'll set up some hidden variables to track our progress in the story. Our five unchecked variables (•*"Drums"*, •*"Guitar"*, •*"Microphone"*, •*"Radio"*, and •*"Key"*) will be used to store and test the progress of the game and will all be ❹ set to 0 at the start of the game. •Set all of them to 0 in a •⚑ event in the *inventory*.

Step 6: Exits and Scene Switching

Before building our game world, we'll need the system used to change scenes. For this we'll need four sprites, each with the **ball** from the Sprite Library as its costume. A fair bit of code will be required to handle the basic action and

Advanced Project 2: Point-and-Click Adventure ◆ 39

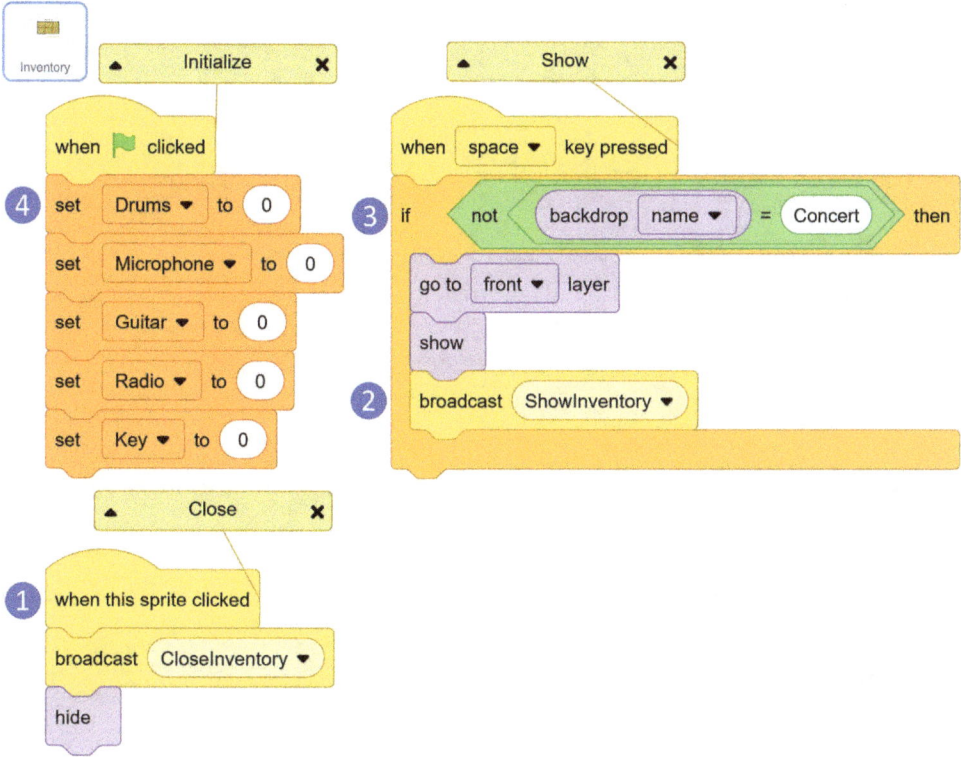

the specifics of our map to allow the right connections from scene to scene. You'll use the exact same principles for all of them, just using different positions and backdrops so we can build a first example, then copy and tweak it for the others.

ExitRight will be our example to explain this system. An exit sprite is a waypoint that the player can see on-screen indicating an opportunity to move to another scene. They'll be able to click on them to move to them, and when they collide into one of these waypoints, it will activate a transition to the next scene. To start, drag our *ExitRight* to its default position, somewhere just below the middle of the right edge of the **Stage Window**. Set its default states in a ❶ •**[When ▷ Clicked]** with a •**[Go To X: (#) Y: (#)]**, •**[Set Size To (50)%]** and •**[Hide]**. Double-check if the •**[Go To X: (#) Y: (#)]** auto-populate has its current coordinates.

The ❷ •**[When This Sprite Clicked]** will include setting the •*"GotoX"* and •*"GotoY"* variables to its position and then •**[Broadcast ["***GoTo***"]]**. When this sprite is clicked, it will get the mouse click event instead of the backdrop, so it needs to cover for what the background usually does in initiating Avery's movement.

Advanced Project 2: Point-and-Click Adventure

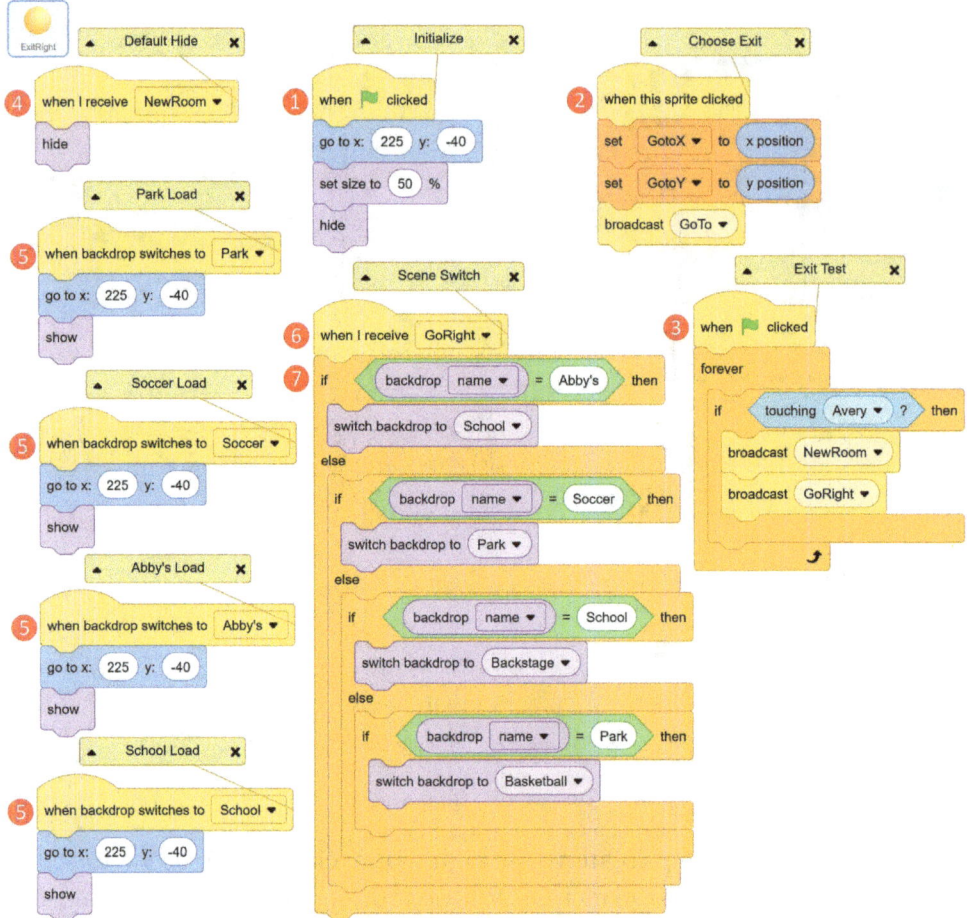

The aforementioned sets the movement to the exit. To handle changing scenes, use ③ a ●[▷]/[Forever] combo to set up a constant condition test. ●[If ●<Touching [Avery]?> Then] will ensure that *Avery* has actually reached the exit before leaving. Inside we will initiate the events ●"NewRoom" and ●"GoRight".

The ④ ●"NewRoom" event will simply ●[Hide] the exit, ensuring it hides by default, so any new scene will only have an *ExitRight* show if it needs one. This means we need to add a ⑤ ●[When Backdrop Switches To [*backdrop*]] event for each scene that has an *ExitRight*. In each of those events, put a copy of the original ●[Go To X: (#) Y: (#)] (with the normal positions) and a ●[Show] block. By doing this, we can go back into each individual scene and adjust the position of the *ExitRight* if needed to match the given backdrop.

Lastly, we need a ⑥ ●"GoRight" **Event**. Chain together as many ●[If <*condition*> **Then** {} **Else** {}] blocks as required to nest in options to determine the exact backdrop the current scene is to determine which destination is correct. ⑦ •[If •<(*backdrop name*) = (*Park*)>] then tests if the scene is the park; inside put a •[**Switch Backdrop to [***Basketball***]**]. In the •**Else {}** stack additional •[**If <***condition***> Then {} Else {}**] to cover each scene that has an *ExitRight*, and set the **backdrop** to the appropriate choice for the next scene to the right. Check the secret map to remind yourself of how the scenes connect.

> **Switching**
> *This project uses a number of very sequence-specific code. You'll see a number of uses of series of If code blocks. Some we chain one under the other; others we chain inside each other, or inside each other's Else {} sections. The difference between these three options is very important to understand if you're going to do advanced projects. A series of independent Ifs will each test things separately and in sequence. If you're testing the backdrop and change it, you could inadvertently cause a chain reaction. [If <(backdrop)=(1) Then {Set (backdrop) to (2)} followed by [If <(backdrop)=(2) Then {Set (backdrop) to (3)} at first might makes sense to switch the backdrop depending on where you are, but both Ifs would run, because the first If changes the backdrop, which then is true for the second If, which would then run and change the backdrop again. You'll never see the backdrop be two in this case. To prevent these kinds of accidental repetitive executions, you can use nesting. In the case of the second If being present in the first If's Else {}, it would only run when the first If has already failed backdrop=1, preventing repetitive chaining. Our third option is nesting Ifs inside the Then {} section. Doing so is the same thing as using an <<> and <>> code block. The second If only runs when the first If is true. The difference is that you could run it before the inner If, when it needs to run regardless of the second If's results. Understandably, it can get a bit confusing at first. Try making a project just to explore these three processes.*

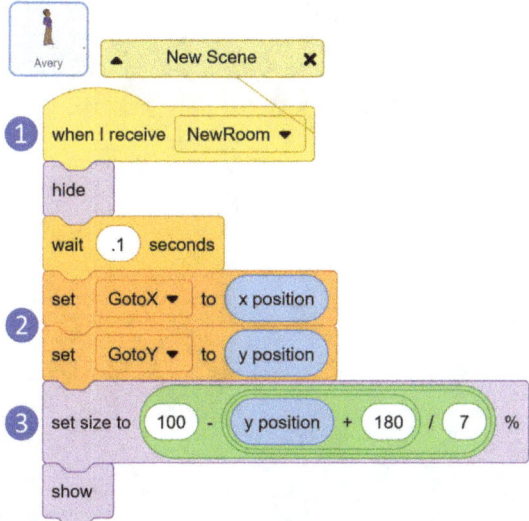

In *Avery*, add a ① •*"NewRoom"* event. Here •[**Hide**] *Avery*, •[**Wait (.01) Seconds**] for the scene to switch. Then set the ② •*"GotoX"* and •*"GotoY"* variables to Avery's X and Y positions. Then copy the perspective scaling combo ③ •[**Set Size To** •((*100*)-•(•(•(Y Position)+(*180*))/(*7*)))) %] so that *Avery* will properly scale to the new scene, and then •[**Show**] to become visible in the new scene.

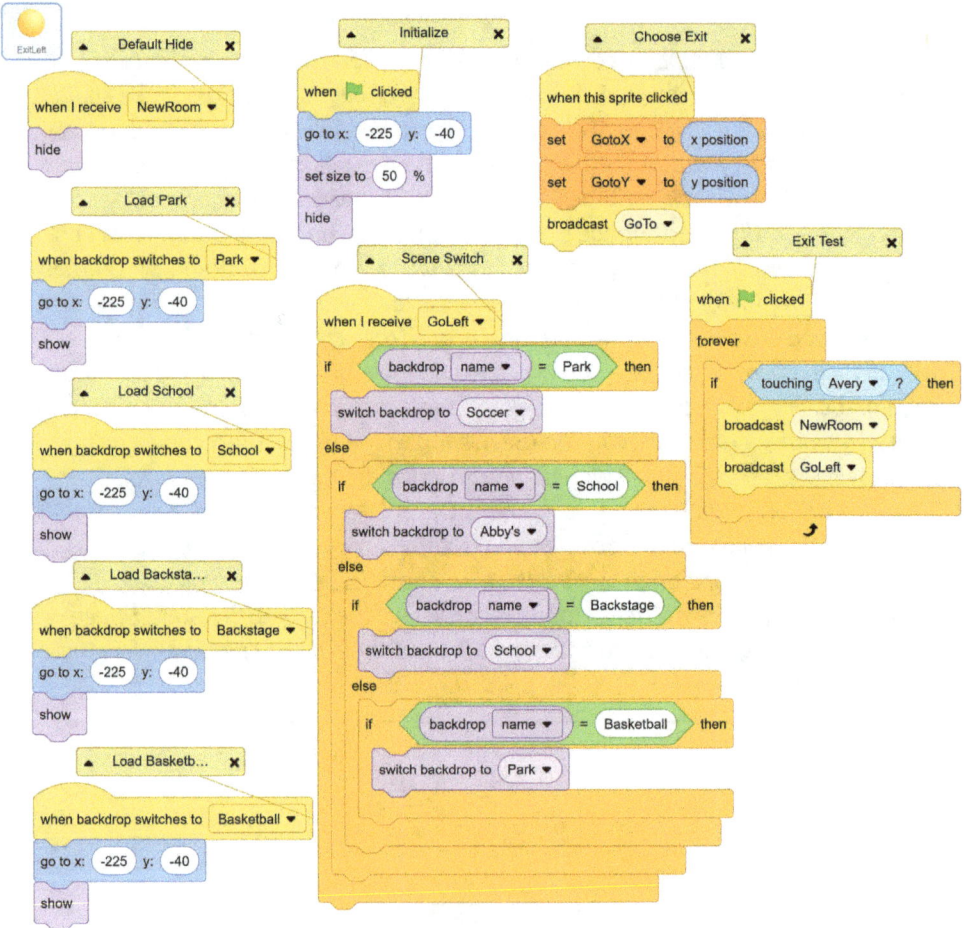

Advanced Project 2: Point-and-Click Adventure ◆ 43

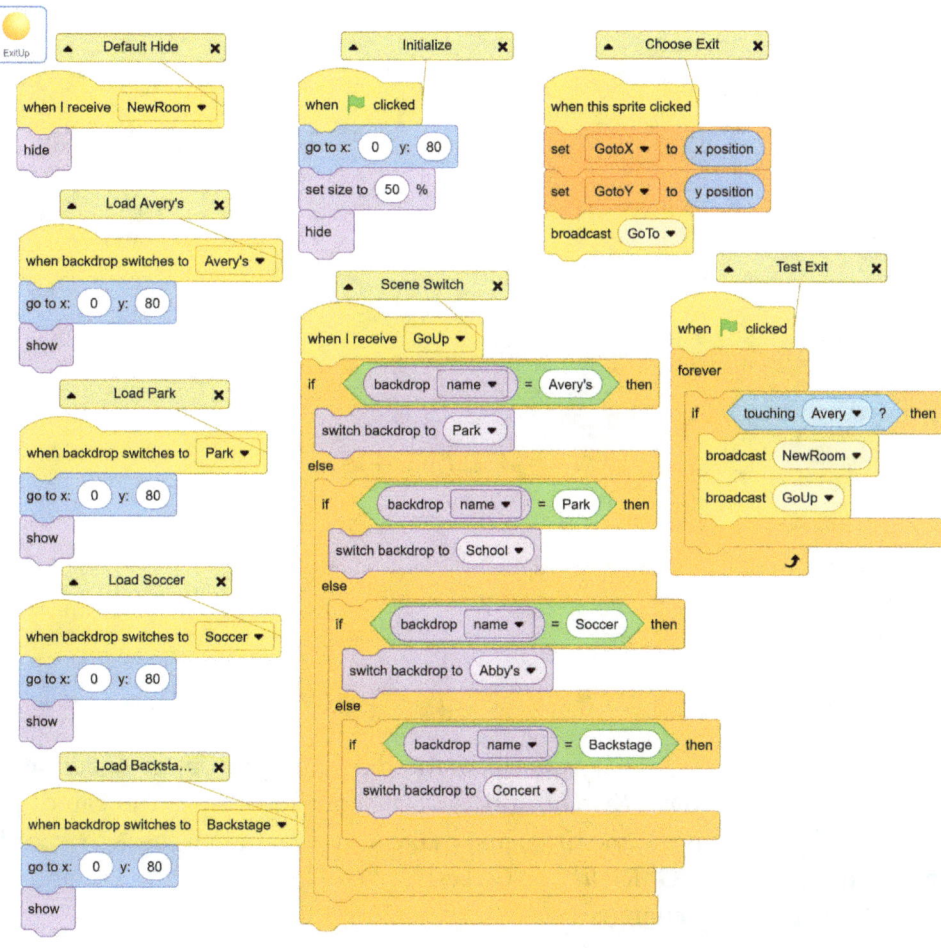

44 ◆ Advanced Project 2: Point-and-Click Adventure

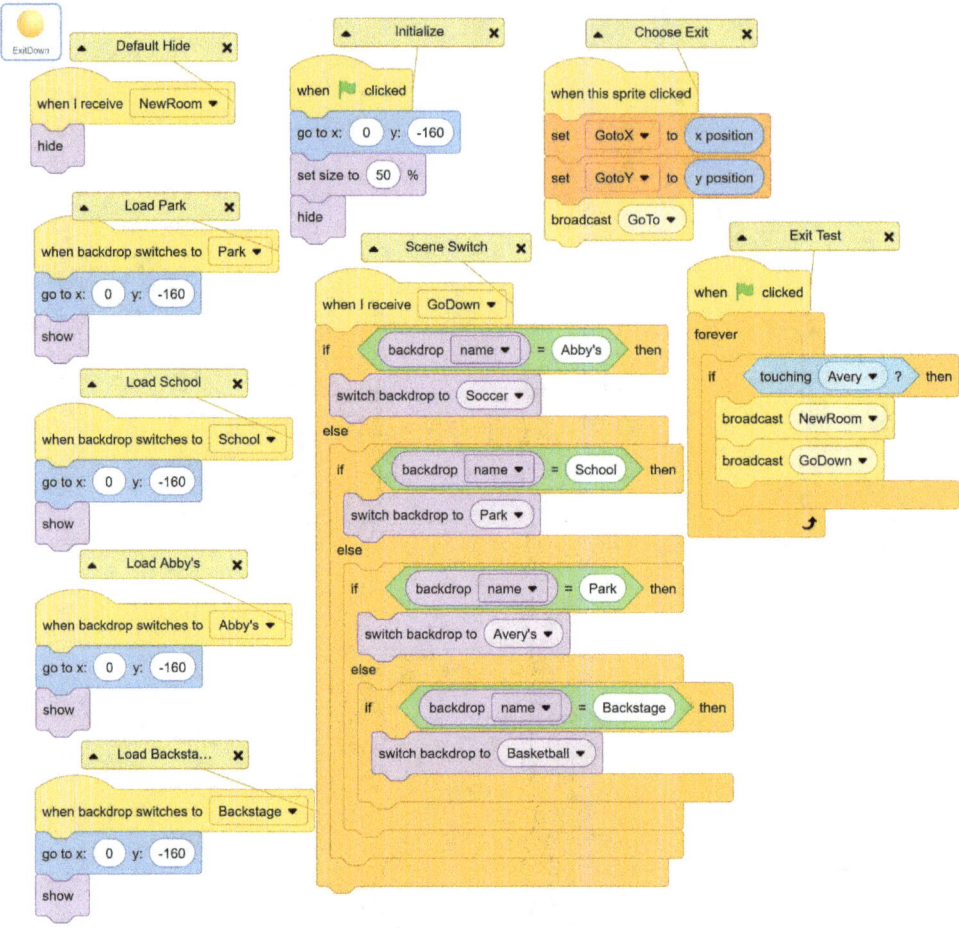

Very importantly, what was done for the *ExitRight* object needs to be repeated for an *ExitLeft*, *ExitUp*, and *ExitDown* sprite as well. You can duplicate the object to save some time; just make sure you get all the X/Y coordinates, the events (●*"GoRight"*, ●*"GoLeft"*, ●*"GoUp"*, ●*"GoDown"*), and the backdrops properly changed.

Step 7: Side Switching

You'll probably notice that the scene transitions work, but there are some issues. You skip over some scenes, and it's a little odd having *Avery* start the new scene in exactly the same position as the last one. It would make more sense to have her walk right and arrive on the left side of the new scene. Let's add some code to achieve that effect, and it will solve our other problem too!

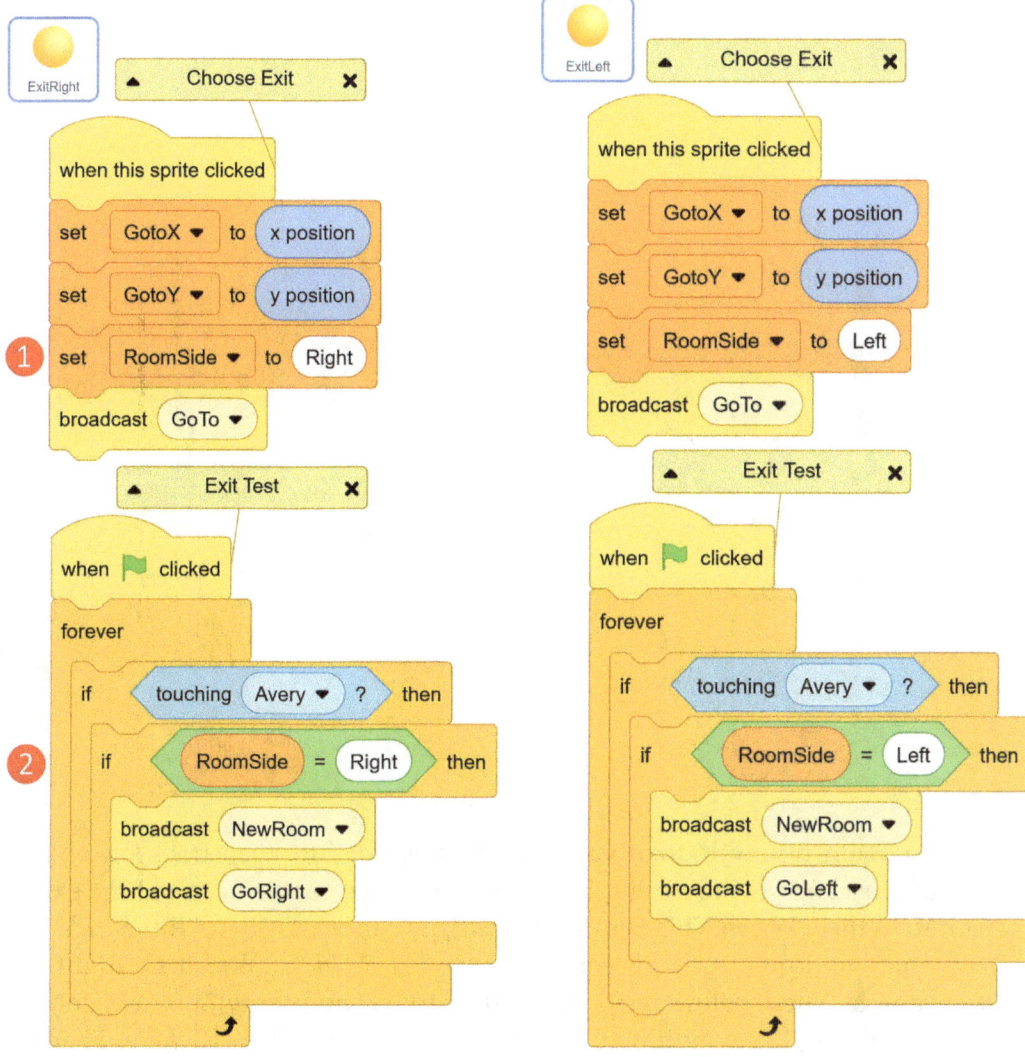

Start with a new hidden variable •*"RoomSide"* to keep track of what side of the screen we're exiting from. Then, edit the code on each of the exit sprites. In the •**[When This Sprite Clicked]**, add in a ❶ •**[Set ["RoomSide"] to (value)]** and give it the direction of exit; *ExitRight* would set to "right", for example. Then, below, inside the •**[If** •**<Touching [***Avery***]?> Then]** code block, add another conditional ❷ •**[If** •**<**•**("***RoomSide***")=(***Right***)> Then]** and put the •*"NewRoom"* and •*"GoRight"* inside it. This will ensure that the exit was actually clicked on before it will register the collision and trigger a scene transition, no more accidentally going up from walking too close below the *ExitUp*. Keep in mind, these changes need to be done on all the exits, with the corresponding code blocks and values for their side.

Advanced Project 2: Point-and-Click Adventure ◆ 47

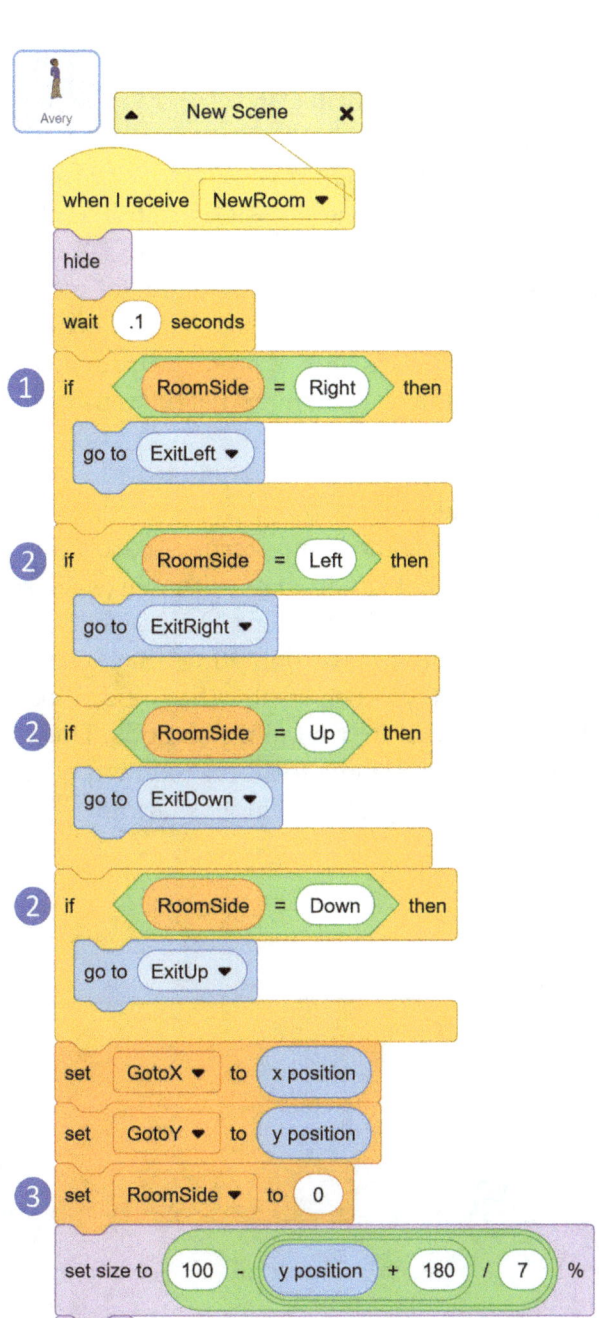

Now, switch to *Avery* to finish implementing the side-switching transitions. In the ●*"NewRoom"* stack above the ●*"GotoX"* setting, we need a series of conditionals; you can make one and duplicate it for the other sides. Each will consist of an ① ●**[If** ●<●**("RoomSide")=("Right")> Then]** with a ●**[Go**

To [*ExitLeft*]]. So here through the ② multiple ifs, you can see we're detecting the •*"RoomSide"* but then positioning *Avery* to the opposite side: right goes to left, up goes to down, and vice versa. We set Avery's position to the opposite side before setting the •*"GotoX"*/•*"GotoY"* variables to the current position so *Avery* is happy in her new position until the player enters a new movement command. Importantly, we also need to ③ •**[Set [*"RoomSide"*] to (0)]** in here. This will help prevent any accidental movement by setting •*"RoomSide"* to a value that no exit will recognize. Only by clicking an exit will it get set to a value that can allow a scene transition. This also prevents Avery from entering a scene and then immediately transitioning back because she's touching an exit.

Step 8: Starting Up and Wandering the World

At this point we'll incorporate the game-starting scene to introduce our story and purpose and go through the map to tweak some of our scenes. Start with *Avery* and add in a ① •*"GameStart"* event. Here, position *Avery*, set her costume, •**[Show]**, and give some exposition with •**[Say]** code blocks. To speed up this step in development, you can create •**[Say]** code blocks that run super quick by just adding a decimal in front of the seconds. This way you can confirm it's all working, but won't have to sit for too long while play-testing as you develop. Remember to correct this before sharing!

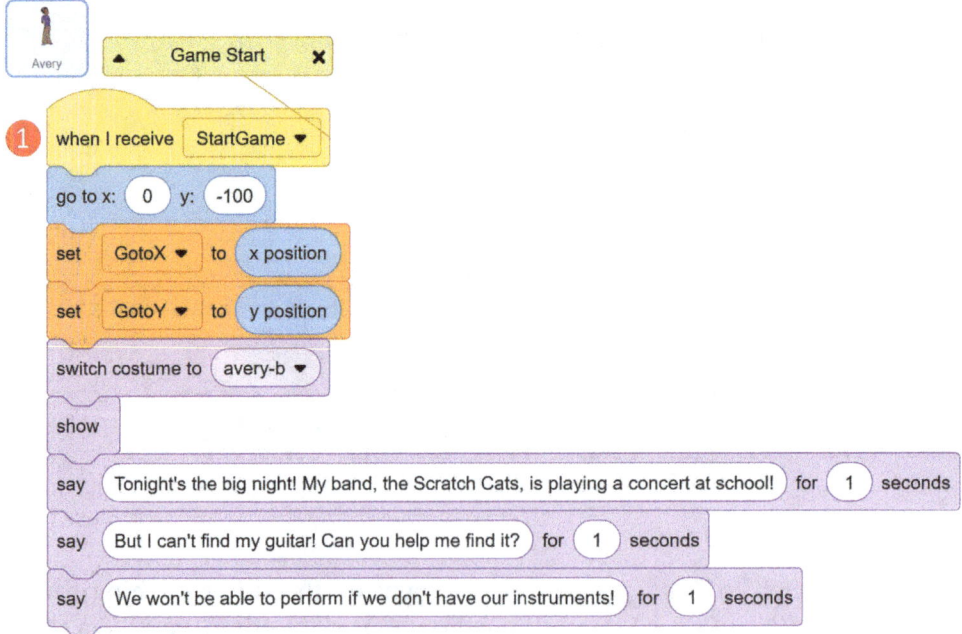

Next, we'll switch over to the *stage*. Now we want to have *Avery* wandering through every map. This is an important play-testing step we can use to correct movement for each map. We'll add a maximum Y value to walk to so that we don't have *Avery* walking into the sky.

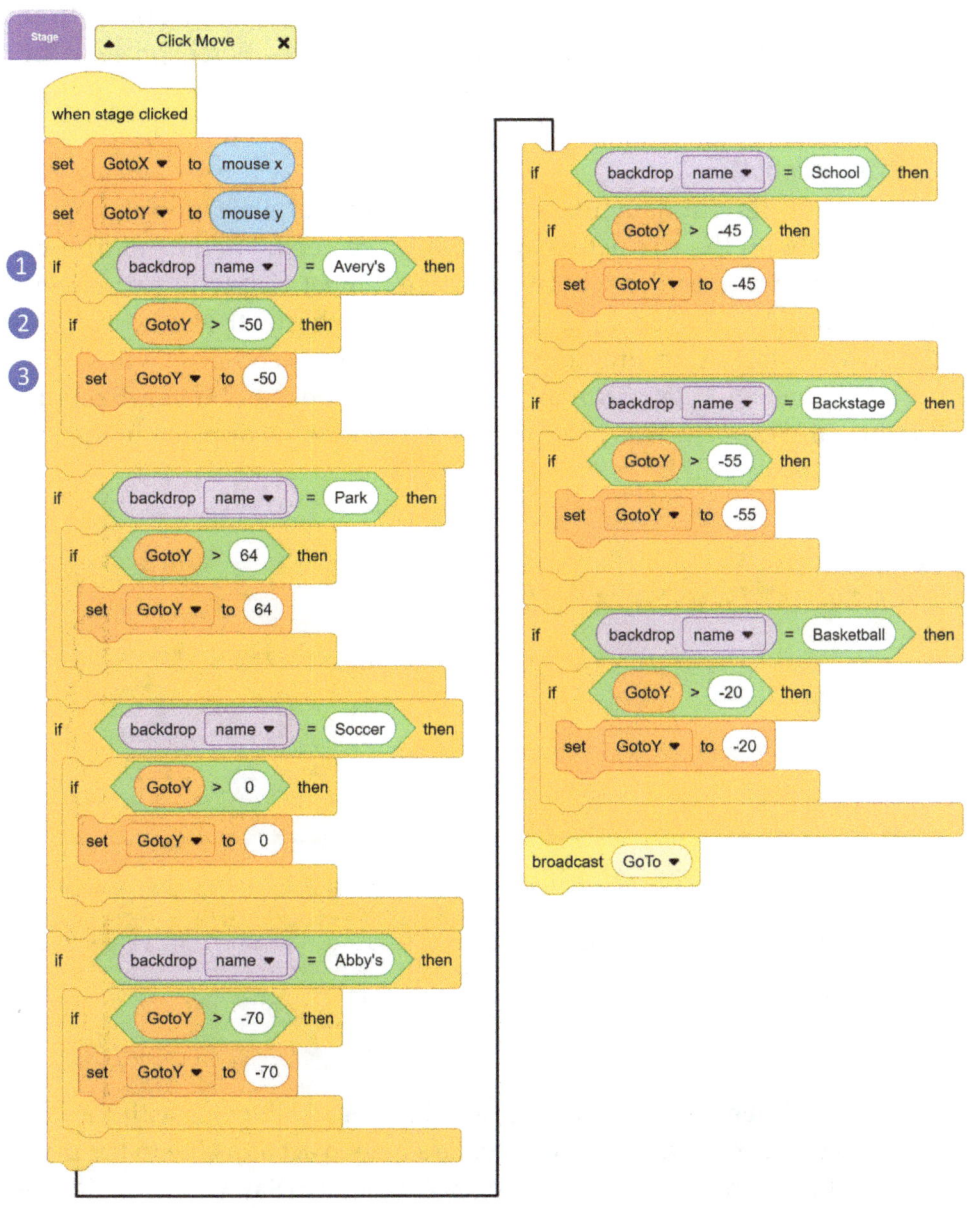

To add maximum Y values to maps, we can move *Avery* to the highest position we would want, making note of that Y value. Go to the *stage*, in the •[**When Stage Clicked**] stack, below the •[**Set** ["*GotoY*"] **to** •(**MouseY**)], but above the •[**Broadcast** ["*GoTo*"]], and add an ① •[**If** •<•(*Backdrop name*) = (*that backdrop*)> **Then**] with an ② •[**If** •<•("*GotoY*") > (*The maximum value you determined*)> **Then**] ③ •[**Set** "*GotoY*" **to** (*the max value*)].

Copy and paste this once for each scene, update the name for that scene, and enter in the maximum Y value wanted for each scene. Now *Avery* will never walk too high in any of the maps. Importantly, you want to make sure that you haven't made your *ExitUp* unreachable, so check if it still works with any limit you've added. Remember, you can change the position of any exit for a specific scene if you want to!

Step 9: School's Out!

Now that all our fundamentals are there, we can start populating our game world with the characters and objects we need. We'll head up twice from Avery's room to the school. Here we'll meet *Kai*, who's discovered that the school is locked up and he can't get the textbook he forgot at school. We'll start by adding the sprite *Kai*. For *Kai's* code, we will •**[Hide]** on ❶ •▷ and ❷ •*"NewRoom"* and •**[Show]** on ❸ •**[When Backdrop Switches to [***School***]]**. This ensures this character only appears in the location we want.

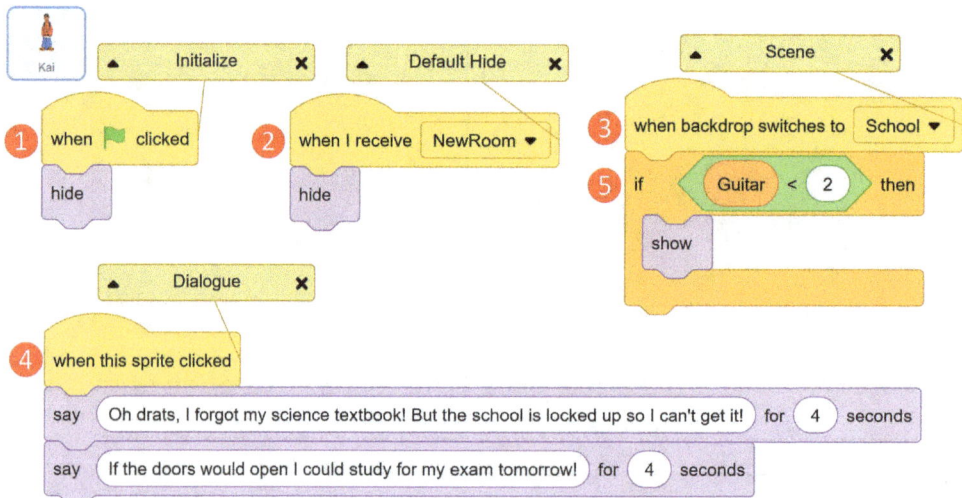

On ❹ •**[When This Sprite Clicked]**, have *Avery* interact with the character. *Kai's* interactions are very simple; he's here to act as a clue to the player that they'll need to get into the school. A couple of •**[Say (*Text*) For (#) Seconds]** code blocks will give our hint to the player to try the school doors. Now, because *Kai* will have his story resolve, we'll actually add in a conditional to his appearance when the backdrop switches to school. You'll need our hidden variable •*"Guitar"* to track the story progress toward finding Avery's guitar. Place the •**[Show]** code block inside an ❺ •**[If** •<•(*"Guitar"*)<(2) >**]** code block. This will ensure Kai only appears if Avery hasn't found her guitar yet.

Step 10: Sasha and Abby

We can duplicate *Kai* to make *Sasha*. We'll switch the **costumes** to *Sasha* for this next character that occurs in the **soccer** scene. Just go to the Costume tab and click Choose a Costume to add Sasha, and after you can delete the Kai sprite from the Costumes List. In the code, be sure to switch the ❶ •**[When Backdrop Switches To [*backdrop*]]** over to **soccer**. In this case, Sasha always shows, so you can just have a •**[Show]** without the •*If* conditional. Sasha will have a much deeper interaction we'll need to code.

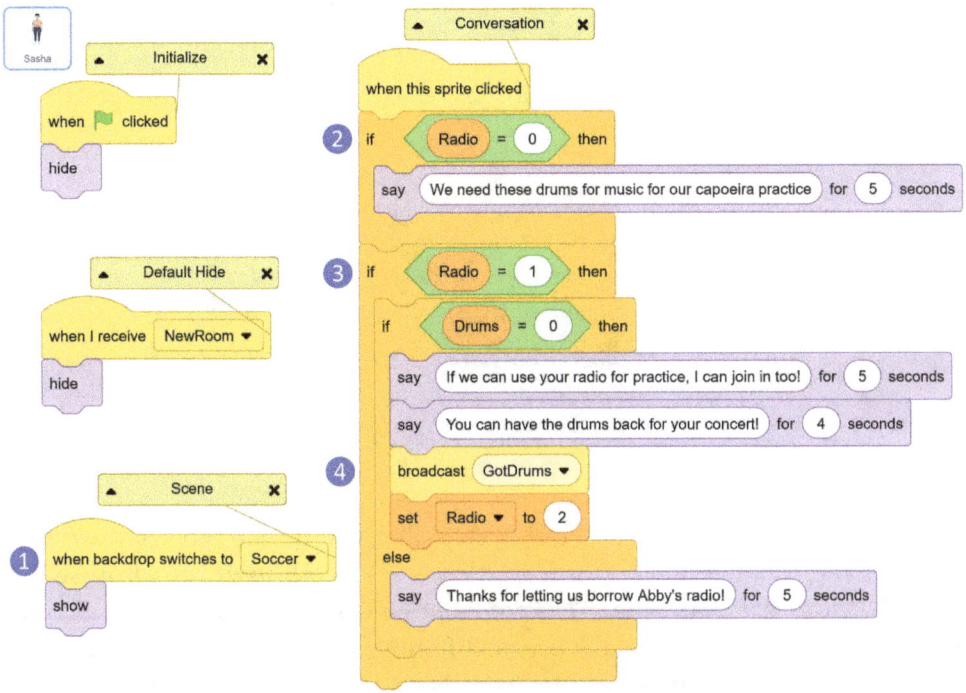

We'll use two different hidden variables here to determine what interaction occurs. *Sasha* uses both the •*"Radio"* and the •*"Drums"* variables to determine events. We'll have a stand-alone If for ❷ •*"Radio"*=0, and then a separate ❸ •*"Radio"*=1. You can set up different logic patterns depending on what changes, or switches, occur in your game, to ensure the right combinations or exclusions are handled in the appropriate manner. Just remember to read through the code for each combination to make sure you haven't missed a state; otherwise, your game might not be able to be played through because an event won't trigger. Missing a combination of conditions could lock you out of being able to run necessary events. We add the ❹ •*"GotDrums"* event to give the player the drums.

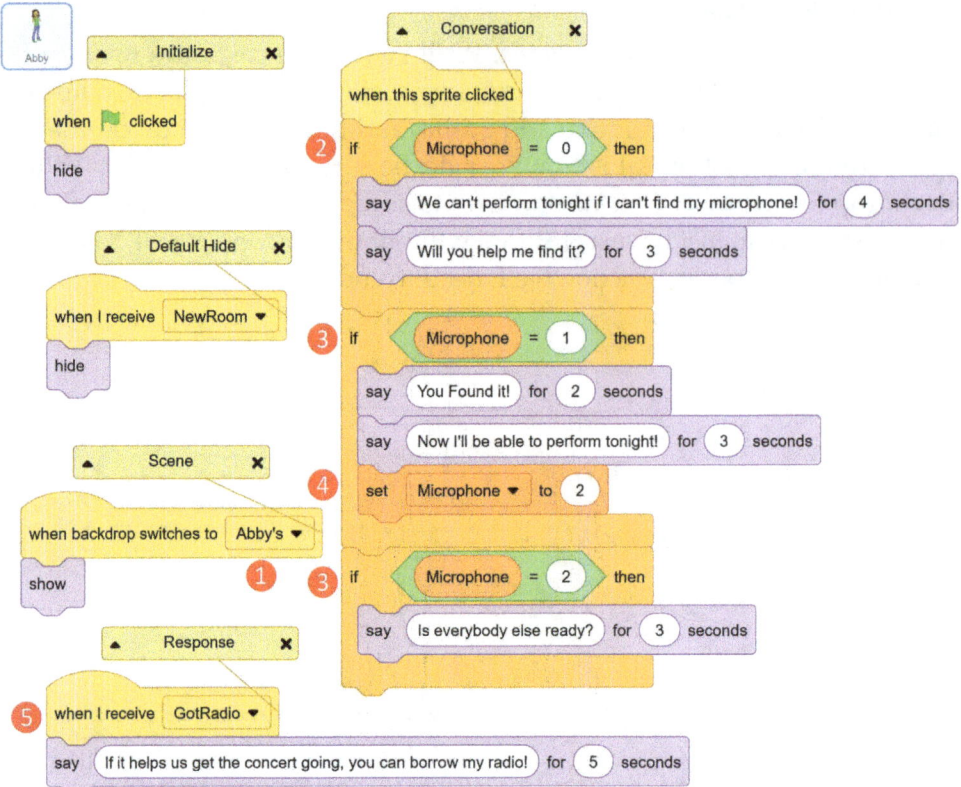

Next, we'll add *Abby*. Copying *Sasha* will give you the most similar code to start. Then change the costume to *Abby* and the •**[Show]** backdrop switch to ❶ *Abby's*. We'll use the •*"Microphone"* hidden variable that will track her story progress, but we'll need a different structure for the conditions. You can delete the bottom two Ifs from Sasha and switch the first one from ❷ •*"Radio"* to •*"Microphone"*. You can then copy that ❸ •**[If** •<•(*"Microphone"*) = (0)>**]** twice to make the three different states that Abby's story can be in. Add in the appropriate •**[Say]** code blocks, and in the •**[If** •<•(*"Microphone"*) = (1)>**]** you'll also need to add in a ❹ •**[Set (***"Microphone"***) to (2)]** for when you successfully return the microphone to Abby.

With *Abby*, add an additional event ❺ •*"GotRadio"*, and add a bit of text with a •**[Say]** code block. Players will be able to interact with an object in *Abby's* room that she'll comment on.

Step 11: The Microphone and Radio

Now we'll need to add our first objects to our game. These represent the items you find in the game world that will help you resolve the different storylines. We'll make the *microphone* first, and then we'll be able to copy it to make our *radio*. Start by adding the *microphone* sprite. We'll resize it, then add a ① •[When ▷ Clicked] event to point the *microphone* in direction 180 so it looks like its lying on the ground, and then •[Hide] it. We'll need a ② •"*NewRoom*" event to •[Hide] it too. Add in a ③ •[When Backdrop Switches to [*Backstage*]] event. This is the scene that the *microphone* can be seen in. Add a conditional •[If •<•("*Microphone*")=(0)> Then] to trigger when the player has not yet acquired the *microphone*. We'll set it to display appropriately in the scene within this •If.

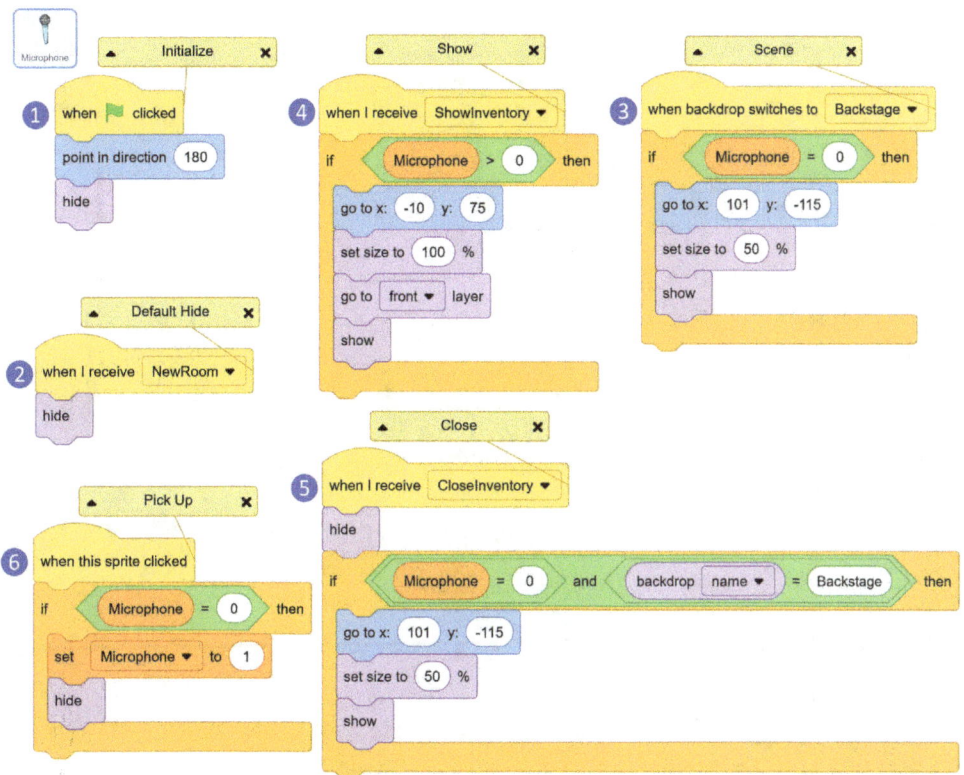

Next, we'll add in a system so the player can view what items they have in their possession – their inventory. We'll need a ④ •"*ShowInventory*" event. This is when the player views their inventory when we'll need to show the microphone in an overlay. The *microphone* should only show up in the inventory if *Avery* has gotten it, so we need a conditional •[If •<•("*Microphone*")>(0)> Then]. Inside we need to position the microphone, size it, and bring it in the front layer so it shows above the *inventory* overlay, and then •[Show]. In a ⑤ •"*CloseInventory*" event, we'll need to •[Hide] the *microphone*, but if *Avery* is in the scene where it is lying on the ground and they haven't taken it, we need to make sure it will show up. So place an •[If •<• •<•("*Microphone*")=(0)> and •<•(Backdrop [*Name*])=("*Backstage*")>>Then]

to test that *Avery* doesn't have the *microphone* and the scene is the backstage. Inside this conditional you can place a ●**[Show]** to ensure it shows up for the player to grab.

We need one more event for the *microphone*. If the ❻ sprite is clicked, a ●**[If** ●<●**("Microphone")=(0)>]** conditional will set the ●*"Microphone"* variable to 1 and ●**[Hide]** allow *Avery* to take the microphone.

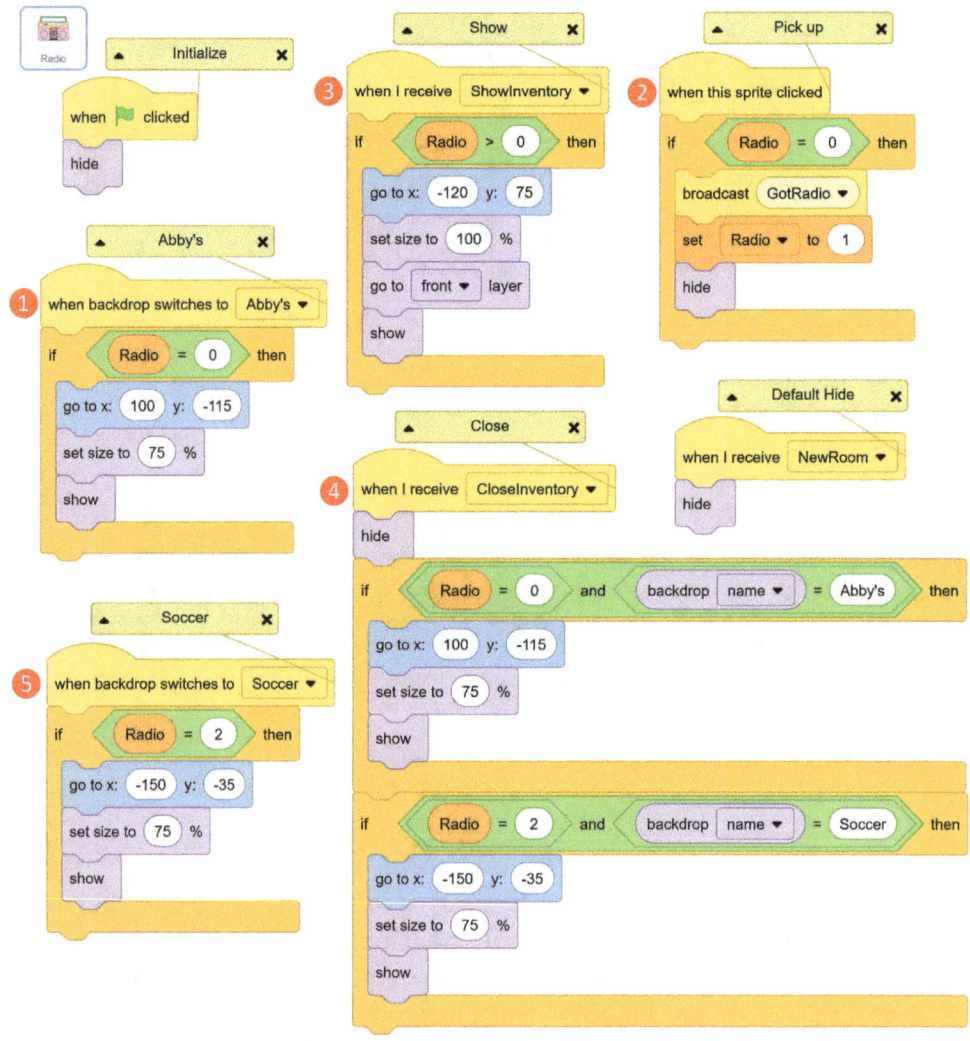

For the *radio*, just duplicate the *microphone* and change its costume in the Costume tab to the *radio*. It should reveal in the ❶ ●*"Abby's"* backdrop event and needs to position correctly there based on a ●*"Radio"* conditional test. In the ❷ ●**[When This Sprite Clicked]** event, it will test for ●*"Radio"*=0 and broadcast a ●*"GotRadio"* event; ●**[Set ["Radio"] to (1)]** and ●**[Hide]** for *Avery* to take it.

You'll need to tweak the ③ •*"ShowInventory"* codes to position it in another inventory block. The ④ •*"CloseInventory"* event will need to test for the •*"Radio"* variable and •*"Abby's"* backdrop. Now we'll also create a second conditional for the •*"CloseInventory"* event. This time, if •*"Radio"*=2 and the backdrop is "soccer", it'll resize, position by *Sasha*, and •**[Show]**. This way, when you have lent it to *Sasha*, you'll see it at the soccer field. You'll also need a ⑤ backdrop switch to soccer to show the *radio*, depending on if you've completed the •*"Radio"* storyline.

Step 12: Noor and the Drumkit

With *Sasha* giving the player the *drumkit*, let's follow that up and resolve the *drumkit/Noor* storyline. Duplicate the *microphone* since it is the most similar object, and switch costume to the *Drums Conga-a*. We'll need to switch the ① backdrop to soccer and the variable to •*"Drums"*, as well as tweak the positioning in the backdrop switch and ② •*"ShowInventory"* events. It will animate on the ③ •*"GotDrums"* event, which is triggered from a state where the drums are already visible and in place, so we can simply start with a •**[Glide 1 Second to X: ([X Position] of [Avery]) Y: (([X Position] of [Avery])+(50))]**, •**[Wait (2) seconds]**, •**[Hide]**, and then •**[Set ["Drums"] To (1)]**. Make sure you tweak the ④ •*"CloseInventory"* event so that if the *inventory* closes, it will •**[Hide]** the *drumkit*, unless you're in the soccer scene and haven't acquired them yet.

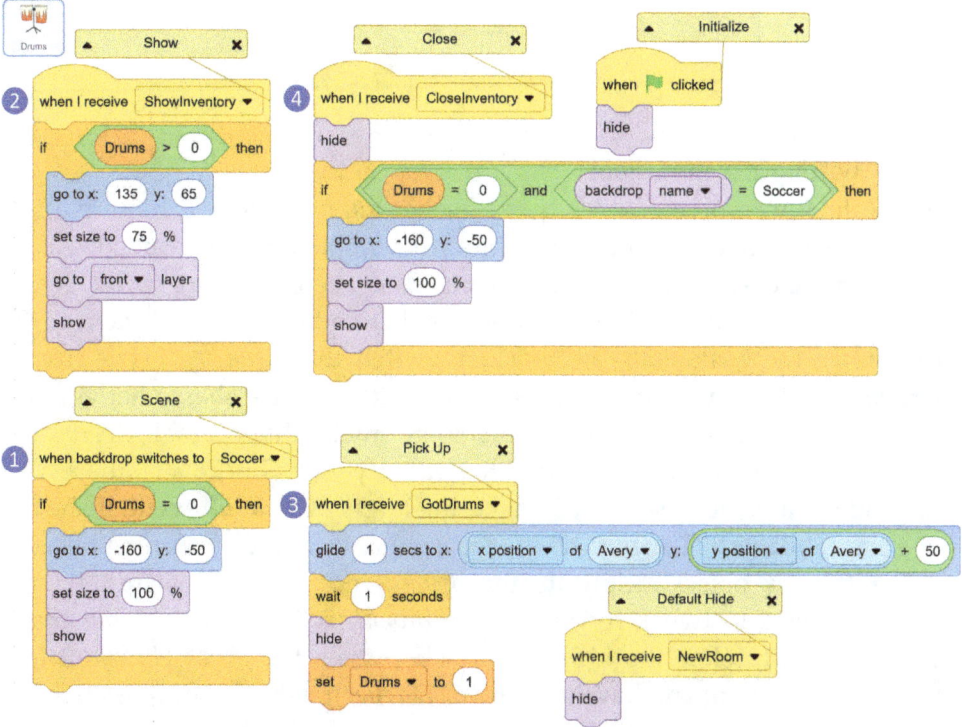

You can copy *Abby* to create *Noor*. Just change the **costume** to *Noor* after duplicating them. *Noor* will ●**[Show]** in the ❶ backstage backdrop. We'll have three possible interactions using the separate stacked sequential ●*Ifs* the way *Abby* had them. Each will include some exposition with say blocks, the ●*"Drums"*=1 ●*If* will ❷ set ●*"Drums"* to 2, and ●*"Drums"* =2 will also call the ❸ ●*"ReadyTest"* event. You can toss the ●*"GotRadio"* event, as there is no equivalent for *Noor*.

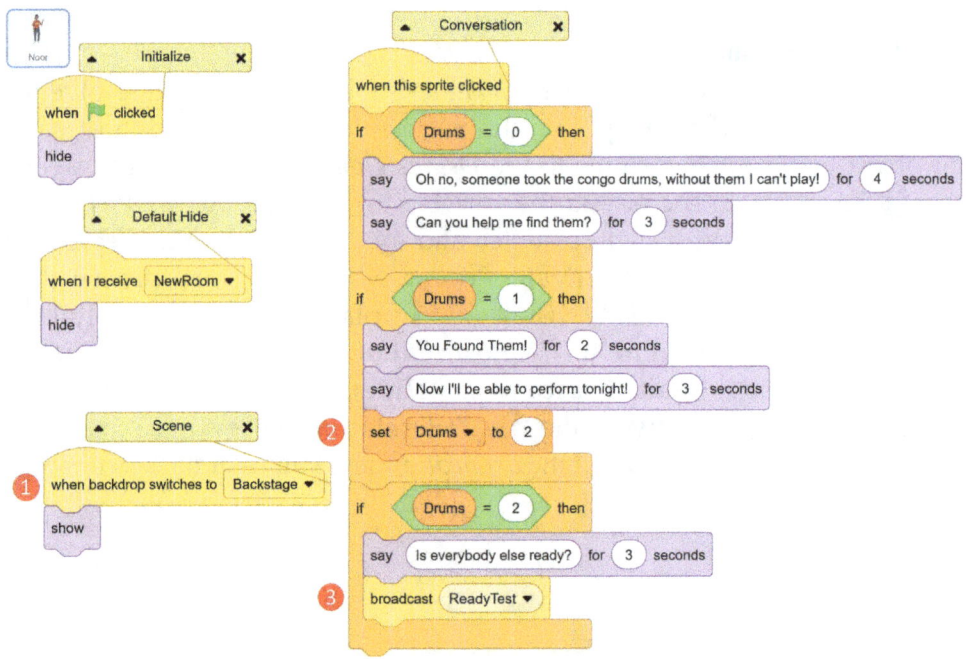

Step 13: Getting into the School

Time to finally resolve Avery's quest for her *guitar*! This will require a few extra sprites. First off, add a sprite for a *teacher* character (we can copy *Noor* and swap the costume). I used a costume from the **characters sprite**, which contains many different character designs. For code, we'll need them to ❶ ●**[Show]** in the basketball scene. Then, switch the three ❷ ●**[If]**s from ●*"Drums"* to ●*"Guitar"* variable tests. If *Avery* hasn't remembered her guitar is in the school (●*"Guitar"*=0), then only some small talk occurs. If *Avery* does remember but she hasn't yet gotten her *guitar* yet (●*"Guitar"*=1), the *teacher* will lend her a *key* for the school, triggering a ❸ ●*"GotKey"* event. If she has her guitar already (●*"Guitar"*=2) the *teacher* thanks her for returning her key.

For the *key* object, duplicate the *drums* and swap its **costume** for a **key**. We'll need to tweak its ❶ ●*"ShowInventory"* and ❷ ●*"CloseInventory"* code and delete the ●*"Soccer"* stack. Switch the //*Pick Up* event to a ❸ ●*"GotKey"* event and have it ●**[Go To [***Teacher***]]** then ❹ ●**[Glide 1 Second**

Advanced Project 2: Point-and-Click Adventure ◆ 57

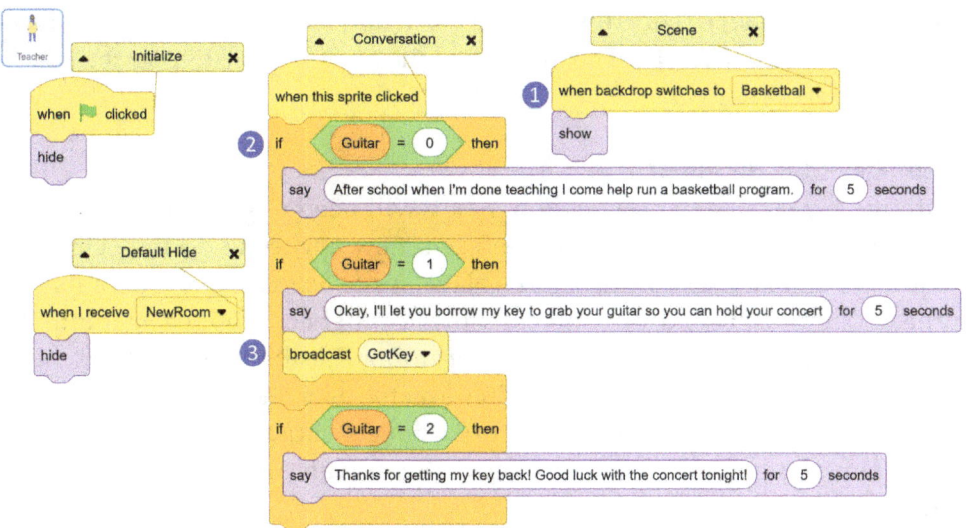

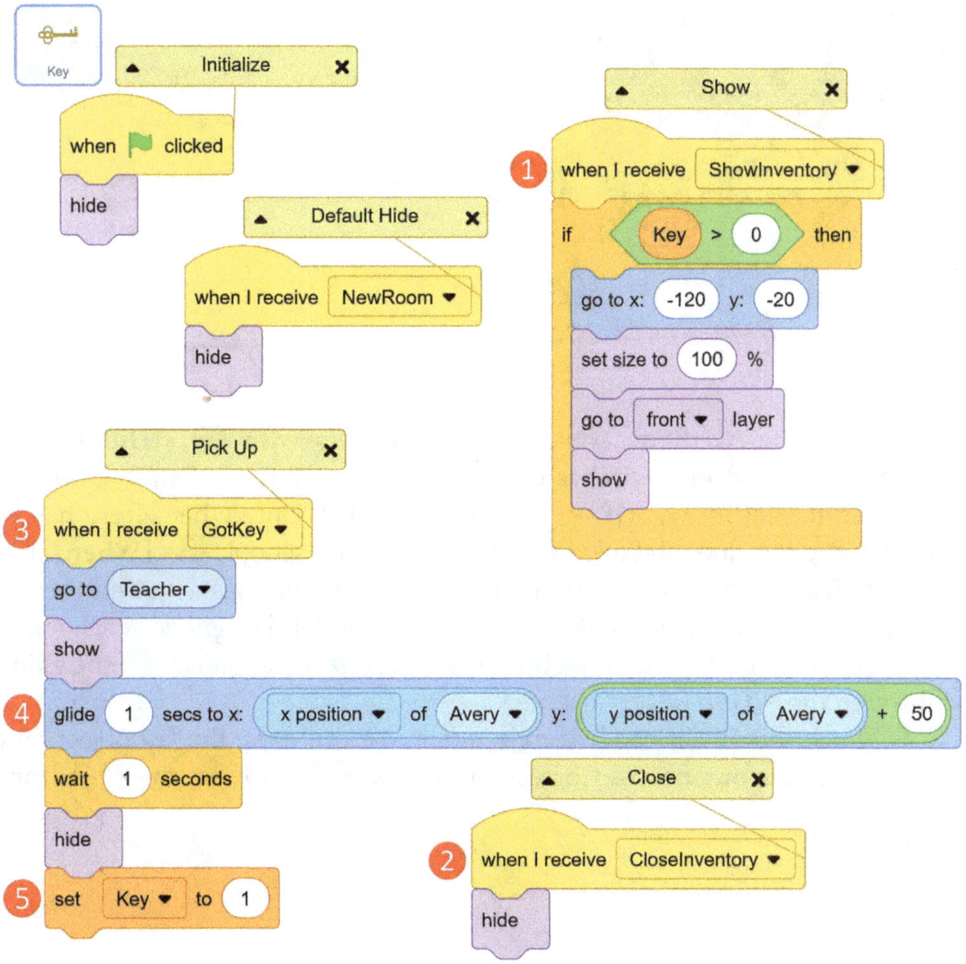

to X: ●([*X Position*] of [*Avery*]) Y: ●(●([*X Position*] of [*Avery*])+(50))] and have it ⑤ ●[Set [*"Key"*] To (1)]. This will allow the *teacher* to give *Avery* the key so she can retrieve her *guitar*.

Step 14: Getting the Guitar

With the *key* now in her possession, *Avery* will need to go to the school and get the guitar from inside. We'll need two new sprites to do this. Our first requires a bit of clever editing. Add a blank sprite *SchoolDoors*, but for its costume, go into the backdrops and copy the **school** backdrop. Paste it into the sprite's blank **costume**. Now, erase everything but the doors of the school. This will give us the doors to position exactly over the background doors, making them look seamless, while providing a clickable object to interact with. This is a great technique to know if you want to make hidden object games.

In the code, we'll do some standard events with ① ●[Hide]. In a ② ●[When Backdrop Switches to [School]] event, have a ●[Show], but we only need it to ●[Show] ●[If ●<●(*"Guitar"*) < (2)>]. Only when *Avery* hasn't gotten the *guitar* does she need to interact with the school. ③ ●[When This Sprite Clicked] holds a conditional to test if the player has acquired the *key* yet. If not, ④ *Avery* gets to *"Remember"* that she left her *guitar* at school. If she has the *key*, has remembered but doesn't yet have her *guitar*, ⑤ she gains it through the ●*"GotGuitar"* event.

To finish this, we'll need *Avery's guitar*. We can copy the *key* object and tweak it to suit our needs. Correct the ① *inventory* to test the ●*"Guitar"*

Advanced Project 2: Point-and-Click Adventure ◆ 59

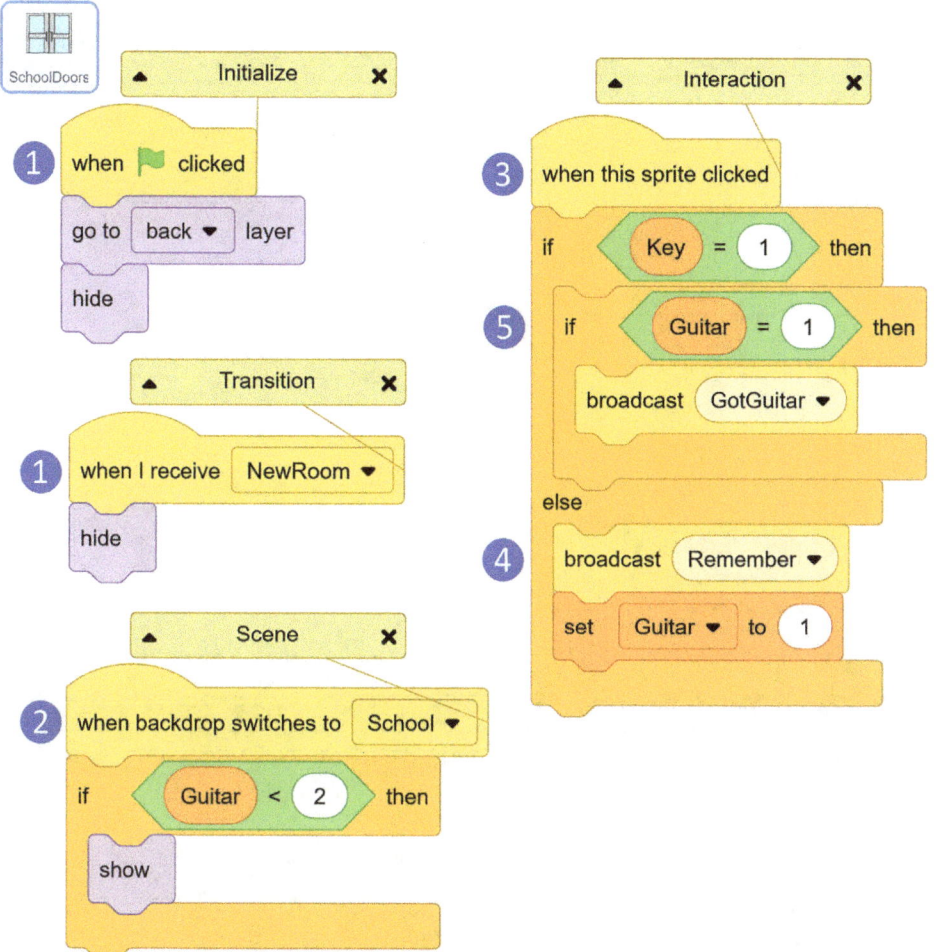

variable, and choose an appropriate **size** and **position** to display it. We'll also need a ② •*"GotGuitar"* event, but we can just adapt the •*"GotKey"* event, switching to a ③ •[Go To [SchoolDoors]], and ④ •[Set [Guitar] to (2)]. In *Avery*, add ① •*"Remember"* and ② •*"GotGuitar"* events with some text to explain what's happening in the game. With these changes, *Avery* will be able to remember her guitar's location and get it when she's gotten the *key*. *Kai* will also celebrate getting into the school; in his code, you can also add a ① •*"GotGuitar"* event, with him thanking Avery for being able to get his textbook before running off.

60 ◆ Advanced Project 2: Point-and-Click Adventure

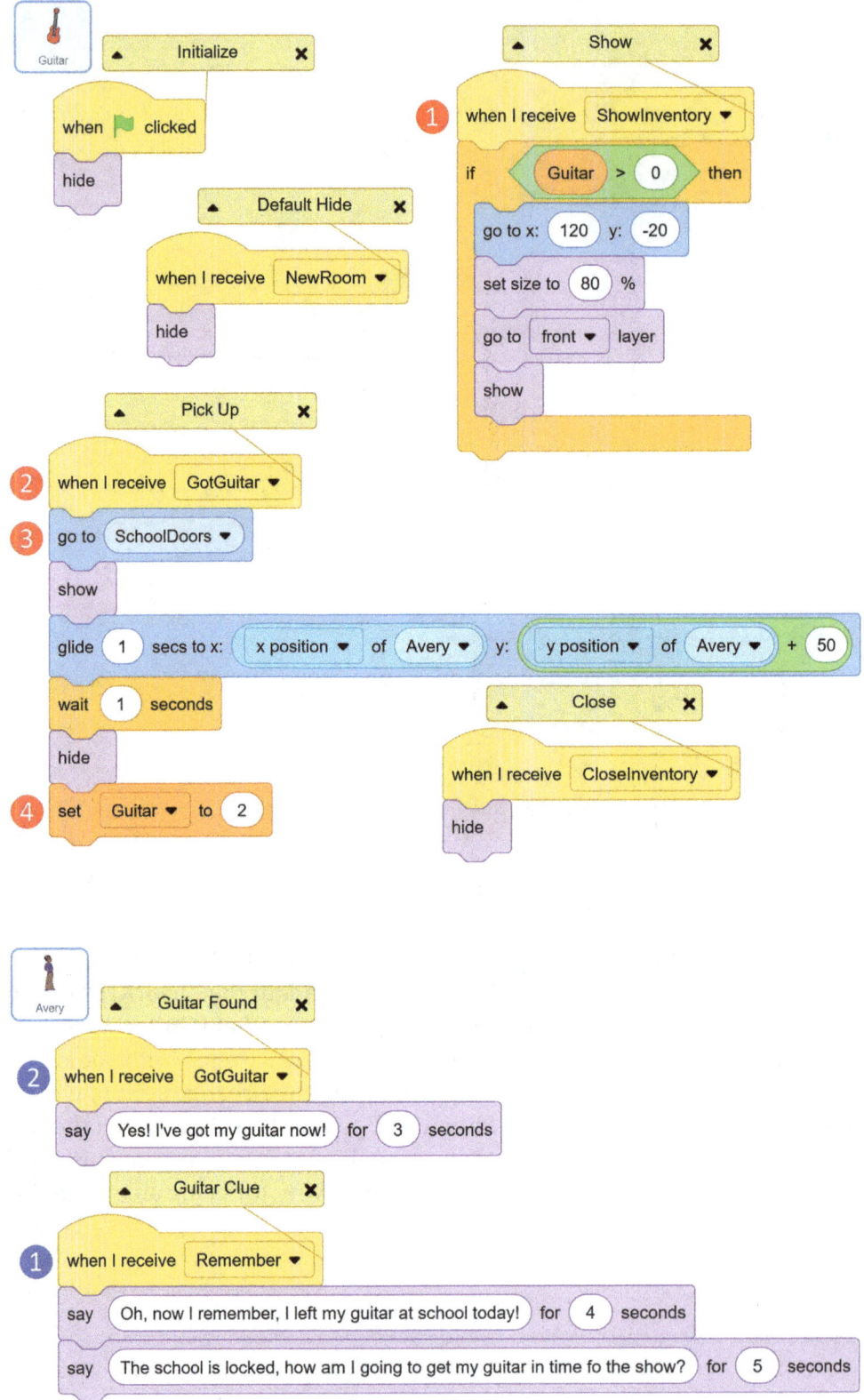

Advanced Project 2: Point-and-Click Adventure ◆ 61

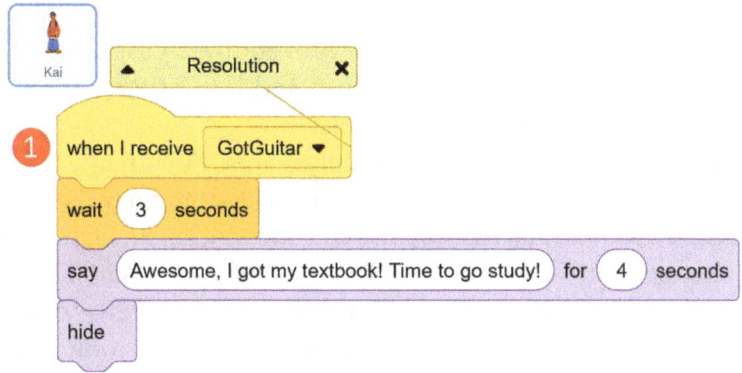

If you're interested in adding in more special effects to celebrate the win, you can check out the Pen Tool Fun project in Book 2: Intermediate for some more ideas and techniques!

Step 15: Ready for the Concert

With all the bandmates' instruments able to be found and acquired, we can finally hold the concert and win the game! But we need to have a way for the code to recognize that the game has been won and be able to proceed on that

basis. Since our final scene is connected up from the backstage scene, we can edit the *ExitUp* sprite to act as a lock. By preventing access until the winning conditions have been met, we create a test, and if its criteria are met, *ExitUp* can appear and allow access to the concert.

We'll need two separate events to handle this. First, make a ❶ •[**When Backdrop Switches To [***Backstage***]]**. Add an •**If** to this event, which will contain the •[**Go To X: (#) Y: (#)**] and •[**Show**] code blocks. Next, to test the various instruments within the same If, we'll need ❷ •[**If** •<•<•(*"Drums"*)=(2))> **And** •<•<•(*"Guitar"*)=(2))> **And** •<•(*"Microphone"*)=(2))>>>> **Then**]. It does look messy, but it works. This will ensure the *ExitUp* only appears if all the instruments have been found.

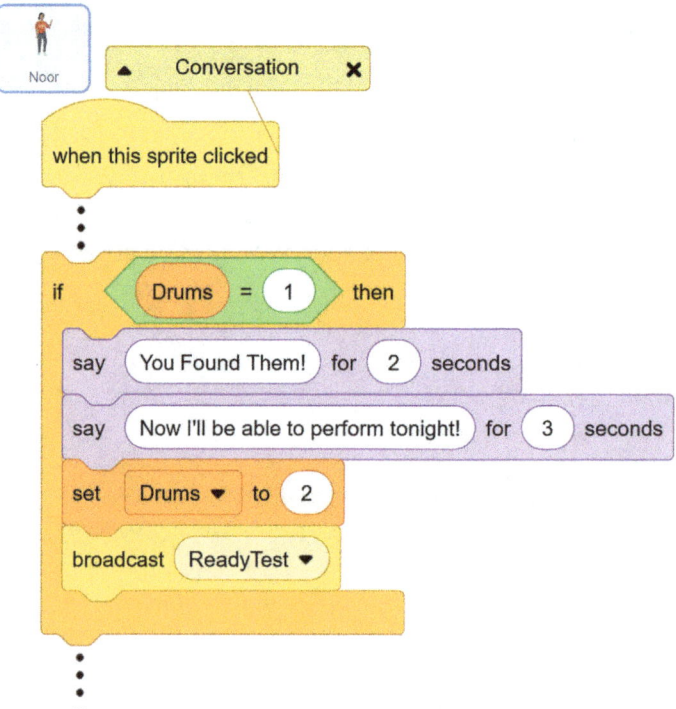

But what if talking to *Noor* is how you get •*"Drums"*=2 and complete all the quests? Well, in *Noor's* code, add a •*"ReadyTest"* event that gets called just after •[**Set [***"Drums"***] to (2)**]. This will allow us to update whether *ExitUp* should •[**Show**] while in the scene, instead of having to leave and re-enter. We can copy our •[**When Backdrop Switches to [***Backstage***]]** stack and simply swap the event for a ❸ •[**When I Receive [***"ReadyTest"***]**].

Step 16: Playing the Concert

Entering the **concert** scene will end the game with one final act. We'll need all the band members and their instruments to appear on-stage for this final scene. Go through ❶ *Avery*, ❶ *Abby*, and ❷ *Noor* and add •**[When Backdrop Switches To [*Concert*]]** events with specific •**[Go To X: (#) Y: (#)]**, •**[Switch Costume To [*Costume*]]**, and •**[Show]** blocks. Then, go through the ❷ *guitar*, ❸ *microphone*, and ❸ *drumkit* objects and add similar •**[When Backdrop Switches to [*Concert*]]** events for them. Optionally, you may want to have the instruments animate to their position to highlight the you-found-them aspect, so you can either use •**[Glide 1 Second To X: (#) Y: (#)]** or just •**[Go To X: (#) Y: (#)]** blocks. You'll need some •**Set Size** and •**Point In Direction** code blocks before •**[Show]** to make sure the instruments look right for the characters.

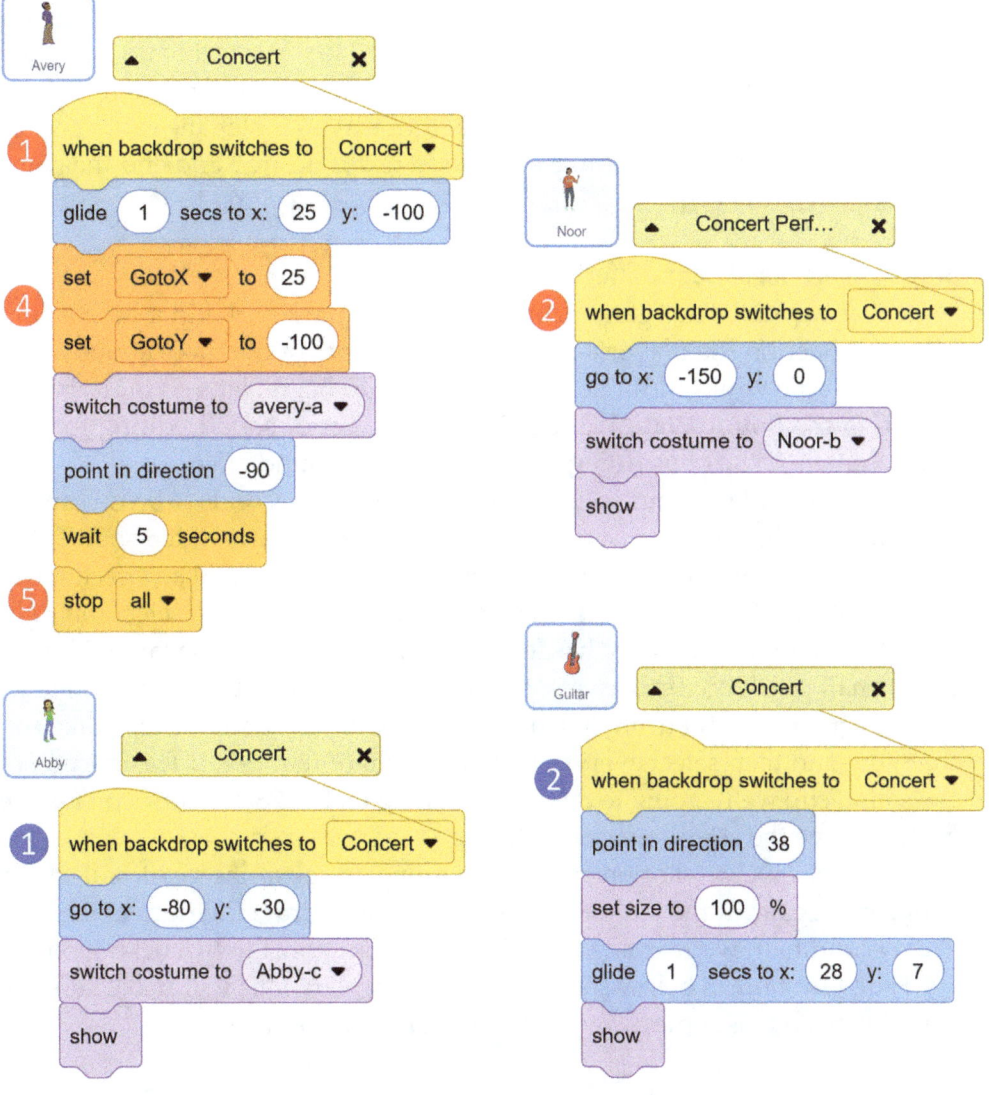

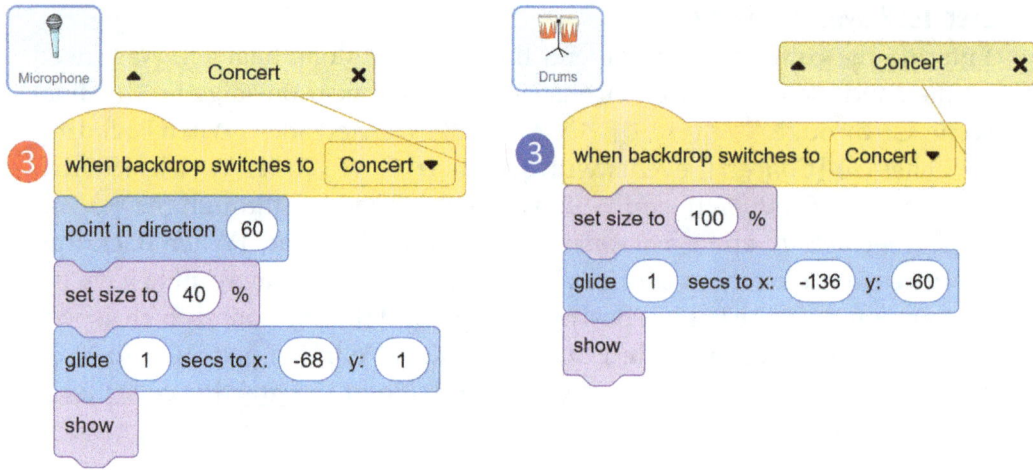

In *Avery*, setting ④ •"*GotoX*" and •"*GotoY*" to the coordinates you want her positioned makes sure she doesn't try to walk around, and using •**Point In Direction** makes sure she's facing the right way to match the *guitar* consistently. You can add in some final exposition text if desired and end the game with a ⑤ •[Stop [*All*]].

Step 17: More about Sound Effects

Now that we know our game works, we can go back and add a little sound effect to cheer on the player's progress. We'll learn a little more about working with audio while adding this sound effect. To begin, let's go to *Avery* and switch to the **Sounds** tab. We're going to do a bit of sound editing for this effect, but we'll start by adding the **Dance Magic** sound effect.

This sound file is a music loop that plays for 8.52 seconds (① you can see a duration in the Sounds List on the left-hand side for each added sound). We just want a short version of it, so we're going to clip it and add an effect. We're going to ② click on the sound wave and drag to the right to select the right half of the sound. The music has four main bars or sections, separated by the three small pinch points between them; we want to select the end of the second pinch point so there are five big spikes on the left, and five big spikes on the right, and then select everything to the right of that point. This portion of the sound clip we're going to delete by clicking on the ③ scissors icon **Delete** button above the sound wave display.

Now that we've shortened it, we're going to select the right half of the sound clip again, at the end of the pinch point, and dragging right to the end. This second half, we're going to add the fade-out effect to by clicking on the ④ **Fade Out** button below the sound wave display. This will make sure the sound fades out as it plays, avoiding any harsh drop-off.

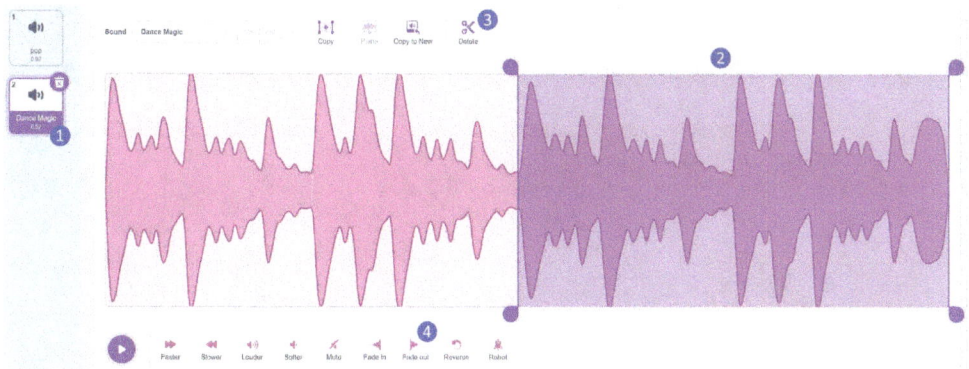

With our sound effect complete, we can now add the code to run it. In Avery's code, we'll add a ① •"*Got Item*" event. In this we're going to add a sound effect as well as play our sound clip. Add a ② •[Change [*Pitch*] Effect By (20)] to make the effect play it faster and at a higher scale to make it even "happier" sounding. Then we'll •[Play Sound (*Dance Magic*) Until Done] and then we'll need to add a ③ •[Clear Sound Effects] block to remove the pitch shift from affecting other sounds. Now, to call this sound effect, we'll go to our *guitar*, *microphone*, *drum*, *radio*, and *key* sprites and call the •"*GotItem*" event when *Avery* acquires the items. It should be paired with the •[Set ["*Guitar*"] to (2)], •[Set ["*Microphone*"] to (1)], •[Set ["*Drums*"] to (1)], •[Set ["*Radio*"] to (1)], and •[Set ["*Key*"] to (1)] code blocks.

Another great touch of sound can be added in – the concert! In Avery, add the sound Bossa Nova. Then at the bottom of the concert stack, just above the •[Stop [*All*]] code block, add in a ④ •[Repeat (5)] with a •[Play Sound (*Bossa Nova*) Until Done] code block. This will give your concert a little taste of music to end with.

If you're looking for more ideas about working with sound and music, you can check out the Dino Dance Party project in Book 1: Beginner!

Step 18: The Main Menu

Lastly, let's polish up the game by adding a main menu and *Play* button. Start by going to our *stage*. Earlier, we added a •"*GameStart*" event under our ① •[When ▷ Clicked]. Delete that because we'll handle it in another object. Instead, after adding it as a new backdrop, add •[Switch Backdrop to [*Spotlight*]]; this backdrop will act as our Main Menu/Title Screen. In the **Costumes** tab, we can click **Convert to Vector**, then add some title text to that backdrop. Then, add a ② •"*GameStart*" event, a •[Broadcast ["*NewRoom*"]], and a • [Switch Backdrop to [*Avery's*]], so when the game begins, it'll set the right scene.

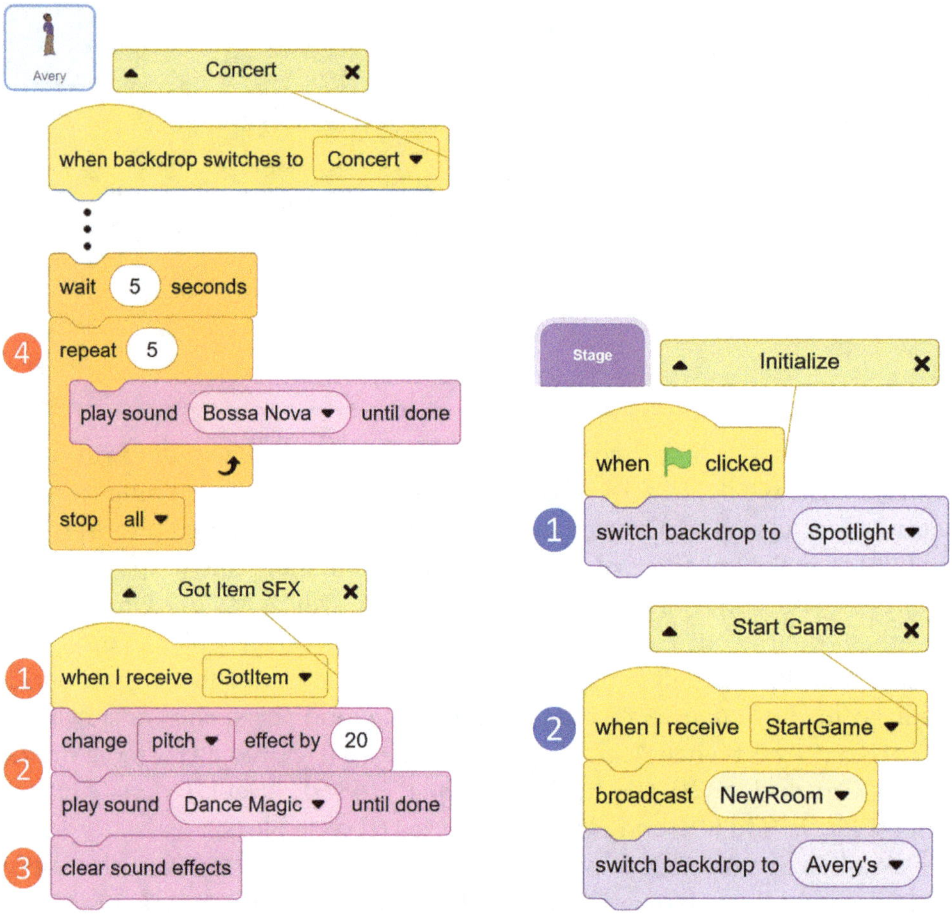

Avery — Concert

```
when backdrop switches to [Concert]
⋮
wait 5 seconds
④ repeat 5
    play sound [Bossa Nova] until done
stop [all]
```

Stage — Initialize

```
when ⚑ clicked
① switch backdrop to [Spotlight]
```

Got Item SFX

```
① when I receive [GotItem]
② change [pitch] effect by 20
   play sound [Dance Magic] until done
③ clear sound effects
```

Start Game

```
② when I receive [StartGame]
   broadcast [NewRoom]
   switch backdrop to [Avery's]
```

You'll need to go back to *Avery* and add a ① •[Hide] to the •⚑ stack to ensure the game starts right. We can't rely on this menu being in front of anything because we used a *stage* costume instead.

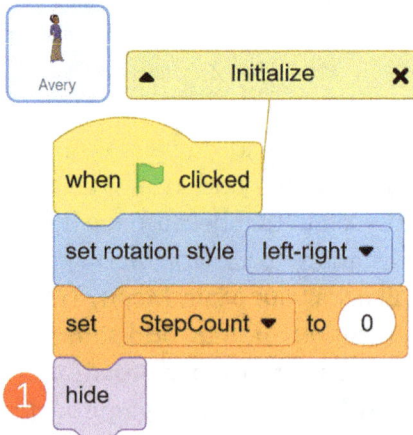

Avery — Initialize

```
when ⚑ clicked
set rotation style [left-right]
set [StepCount] to 0
① hide
```

Let's add a *Play* button to get the game started. Add the **Button 2** sprite from the Sprite Library, and in the **Costumes** tab, add in ① text for the button to say "Play". I prefer using the ② second orange costume for the *Play* button because it stands out more against the **spotlight** backdrop. Click on the costume you want and drag to position the button. Since we won't be changing its position or costume, these defaults will be fine without coding that in.

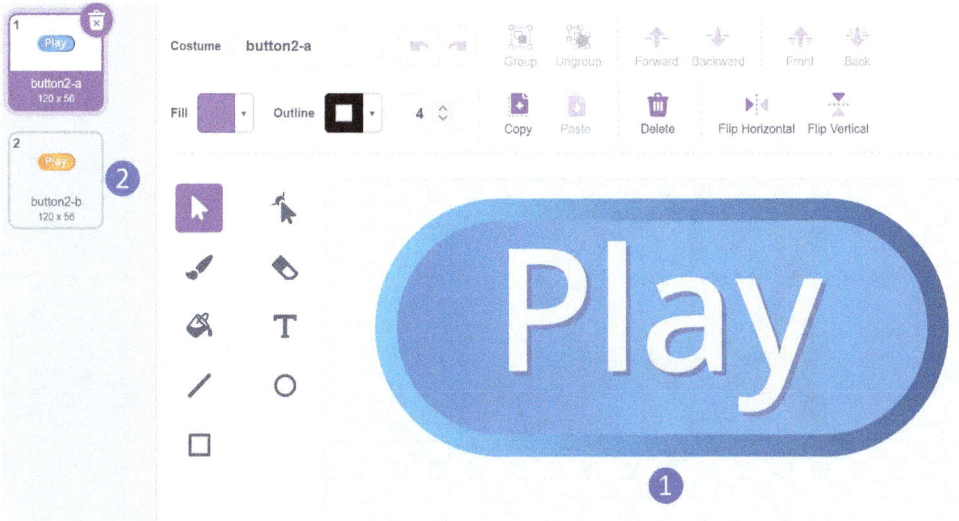

Our code for the *Play* button is mainly for show. We'll animate it to entice the player to click it. In a ① •[When ▷ Clicked], we will set it to •[Point in Direction (90)] and •[Show] the button, then start a •[Forever] loop. Just •[Turn Right (5) Degrees] •[Wait (0.2) Seconds], then turn back, [Wait (0.2) Seconds] then •[Turn Left (5) Degrees] •[Wait (0.2) Seconds], and then turn back and [Wait (0.2) Seconds]. This will create an animation that has the button wiggle left and right while the player is on the menu screen. ② •[When This Sprite Clicked] will •[Wait (0.5) Seconds], run a •[Broadcast ["*Game-Start*"]], and then a •[Hide]. So when clicked, it disappears, and the game begins!

You've now completed your first advanced game project and learned a bunch of handy techniques for building your own adventures!

For working copies of this and every project in the book series, visit www.massivelearning.net for direct links to Scratch Projects, and to see our other projects and resources for coding education!

68 ◆ Advanced Project 2: Point-and-Click Adventure

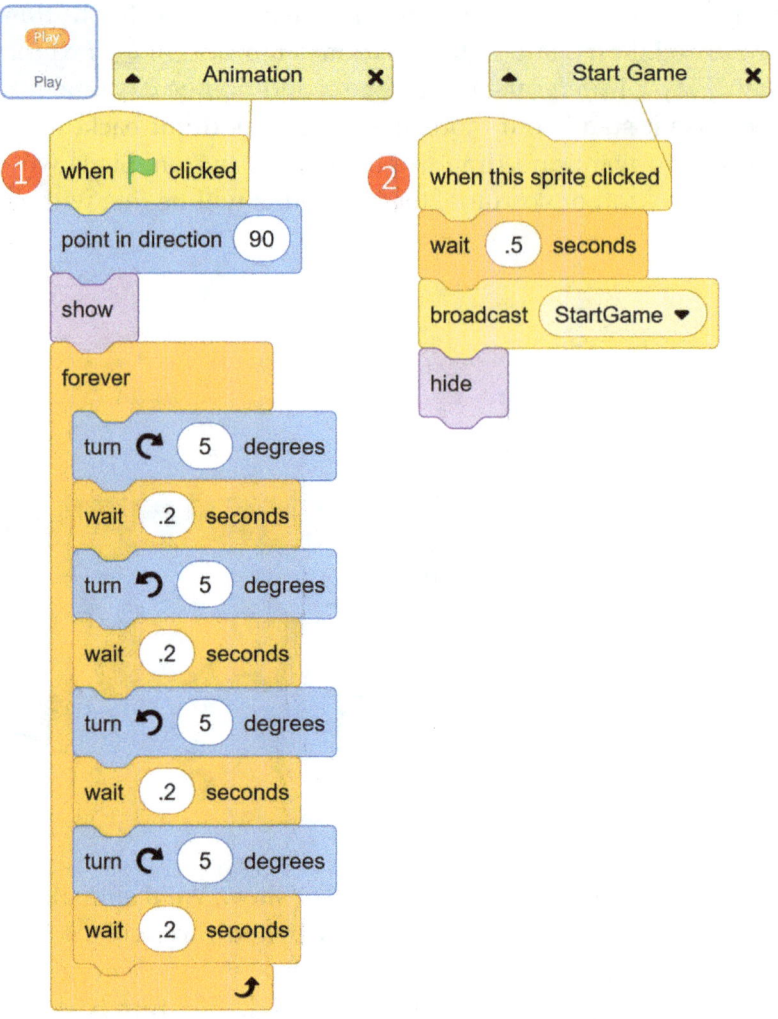

7

Advanced Project 3: Platformer Game

What This Project Is

This project is perhaps the quintessential video game for most kids. A platformer game has players controlling a character that can run and jump around the level, generally a side-on 2D area, with multiple platforms to move around and between. These games focus on exploration and acrobatics, with players learning to use timely and accurate movement to negotiate their way through challenging courses. With multiple levels to work through, we'll see lots of classic mechanics from platformers which will provide a wide range of possibilities for designing interesting and challenging levels. Players will need to get to the waypoint flag to unlock the final exit, but they can lose a life to spikes or falling off-screen. In around two hours (for adults), we'll build our basic platformer, though it gets tempting to add in more levels to extend the project.

What We're Learning with It

A platformer game allows us to tap into kids' common desire to make video games and explore. Your students will almost certainly ask you about making one, so getting some exposure to methods to deal with some of its trickier aspects will be handy for giving some tips and hints. As this is an action game that's focused on individual characters, we'll learn methods for handling rapid and fluid movement, including running and jumping, with expressive

70 ◆ Advanced Project 3: Platformer Game

animations. We'll also see a number of common objects for platformers, including spikes, pits, waypoints, ladders, moving platforms, and blinking platforms, and see how they can be created through code. A level progression system will take the character from one level to the next and finish the game at the end. This project is a great starter for game design, providing lots of tools for students to create their own levels and games using the techniques. They won't just remix a popular platformer project but build it from scratch to understand it while still getting the fun they want.

> **Collision Masks.** A common game development technique for testing collisions for an object using something other than its displayed graphics; a bit of a stretch for Scratch, but a powerful and handy technique.
> **Collision-Tested Movement.** Our movement system will ensure the character doesn't move through walls or fall through floors but also so they can't jump into walls or up through floors.
> **Jumping Movement.** A good technique for jumping movement, with tests for collisions and special terrain for power jumps.
> **Slippy Movement.** A simple function to add a little inertia to the character's movement.
> **Movement State Graphics.** Dynamically draw the character's sprite based on the state of movement and health, giving a basic example of state machine graphics.
> **Death Transitions.** Use of fader transitions for level switching, but also to highlight player deaths and add a dynamic player reset based on level progression.

Terrain Interactions. A number of different obstacles and opportunities for the player to interact with in the level design, allowing them to move in different ways or otherwise be affected.
Waypoints/Progress Markers. Add in a waypoint object that will save a player's progress through the level.
Level Cloning vs. Level Costumes. Two different methods for handling level switching, so you learn when/where to use each.
Inherited Movement. Allow the user to be pushed around by level objects so you can see how we can have the player inherit movement from other objects.
Player Lives and Restarting. Limit the player to a set number of lives and reset their progress based on deaths.

Building It

Step 0: Create Your New Project
Make sure you're logged in to Scratch, then click Create to begin a new project! Since we won't be using it, we can delete the Scratch Cat sprite by clicking on the trash bin on that sprite's thumbnail in the Sprite Listing.

Step 1: Ground to Stand On
With this as a side-view-perspective gravity-based game, we want to make something characters can stand on before adding a character. Add a blank sprite *ground*, and in the **Costumes** tab, create a few levels worth of terrain. For those unfamiliar, a level is a distinct area or challenge that is played in a video game. Each costume will be a distinct level with its own layout, challenge, and goal. You don't have to create the exact same level designs I did, but you'll want a combination of solid ground, pits, platforms, platforms you can hit your head on, difficult jumps, vertical walls, and unjumpable gaps. The examples are very simple, single-colour rectangle-based terrain, but you could make it look nicer. Do be careful to avoid slopes, since we won't be getting into how to program walking up and down slopes (it's complicated) in this project.

For code, we'll keep things simple with a ❶ •**[When ▷ Clicked]** and •**[Go To [Back] Layer]** to make sure it's displayed behind everything else.

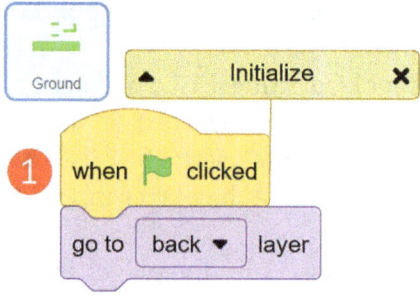

The Pen Tool Fun project in Book 2: Intermediate has some great skills to learn that you could combine here to make dynamic terrain or special effects!

Step 2: Our Moving Player

Now that we've got something to stand on, we can create our character. Add the **Dinosaur4** sprite, but rename it *MovePlayer*, because this sprite will control how our player moves. We want the object to just flip left–right but not rotate, so we'll ❶ •**[Set Rotation Style to [left-right]]**. To make sure it fits the scene, ❷ set its **size** property to 35%. For controls, we'll start with just ❸ left and right, letting the player use either the arrow keys or WASD keys.

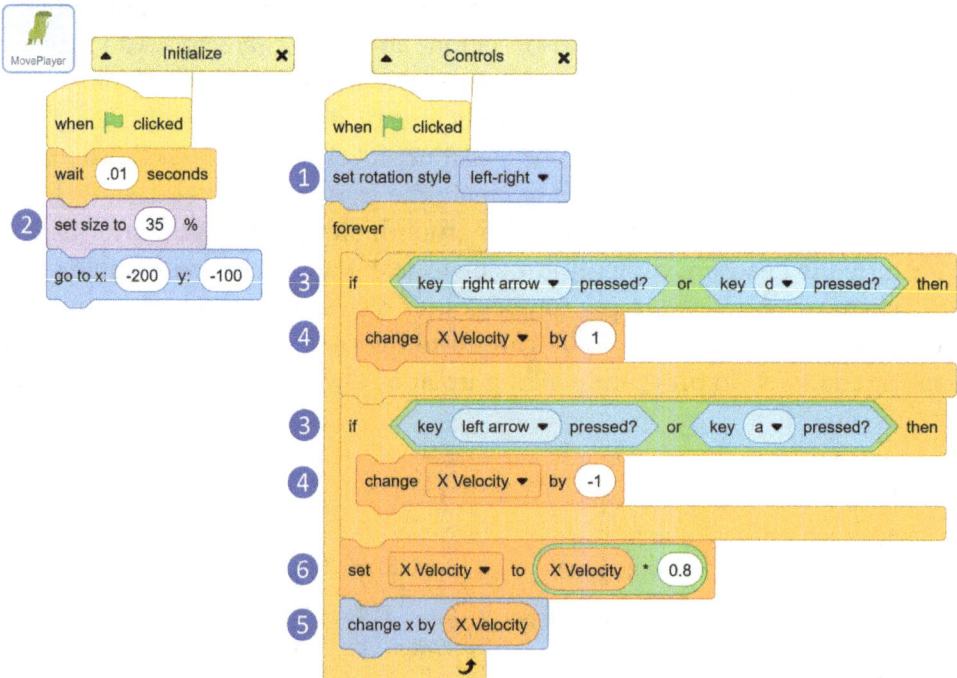

Instead of moving a set amount with a move step block, we're just going to affect the character's speed. Create a hidden •"X Velocity" variable to track how fast the character should be moving. ❹ A negative number on X moves left, and a positive moves right when we ❺ •[Change X by •("X Velocity")]. To ensure our character doesn't move forever, ❻ we slow down the character each step. This resistance or wind down will give weight to our character's movement: the more they move, the more time (and space) it takes for them to slow down and stop. You'll notice that, at the moment, you can move through walls or across any gaps in your level.

Step 3: Collision Testing

To ensure our character can't walk through walls, we employ collision testing. To do this, use a custom •My Block to keep our different processes streamlined and separate. This helps keep our code easier to read. Make a ❶ •"XCollisionCorrection" block and call it under our •[Change X to •("X Velocity")] combo.

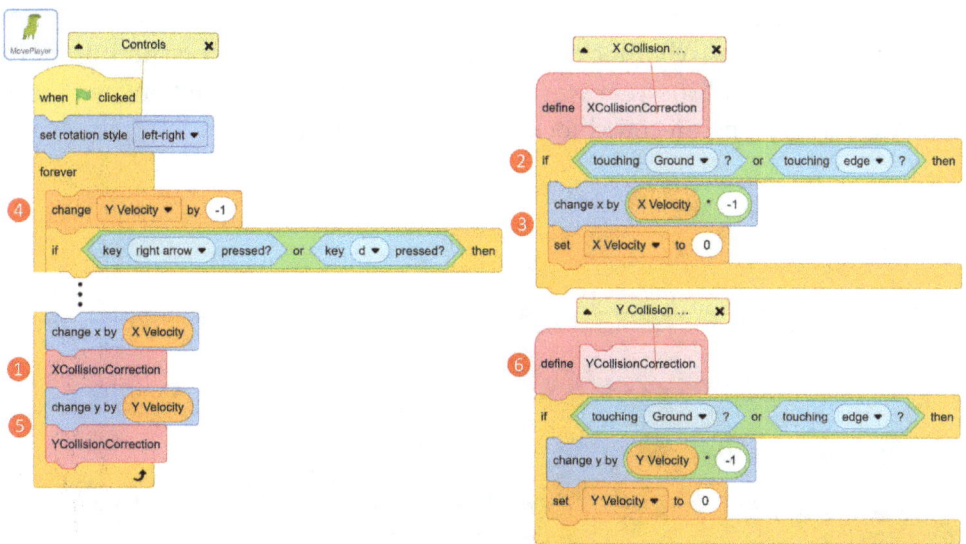

Our •"XCollisionCorrection" custom block will be used to test if the character has run into a barrier, either a portion of the *ground* costume like a wall or platform or the edge of the Stage Window. To do this, use a conditional. ❷ •[If •<•<Touching (*Ground*)?> or •<Touching (*Edge*)?>> Then] allows us to use either the *ground* or edge to stop the character. Inside it have a ❸ •[Change X by •(•("X Velocity") * (-1))] and •[Set ["X Velocity"] to (0)]. This not only stops the character but also moves it back to where they didn't collide and could freely move (otherwise, they'd be stuck forever). By •[Change X by •(•("X Velocity") * (-1))], we are inverting the direction.

We multiply •*"X Velocity"* by -1, which turns a -1 speed into a +1 speed, or vice versa. By inverting the •*"X Velocity"*, we ensure the colliding character moves back the opposite direction and the exact same amount they moved to trigger the collision.

With our •*"XCollisionCorrection"* working, start implementing a gravity system for our game. Add a hidden •*"Y Velocity"* variable and a •*"YCollisionCorrection"* custom •My Block. To implement gravity, add a ④ •**[Change ["Y Velocity"] by (-1)]** block above our movement conditionals. This will make the character always fall when we add in a ⑤ •**[Change Y by** •**("Y Velocity")]** and •*"YCollisionCorrection"* call at the bottom of the •**[Forever]** loop. Every step of the game, the character is now pulled down by gravity. If we copy the •*"XCollisionCorrection"* stack and place it into the ⑥ •*"YCollisionCorrection"* event and swap the •*"X Velocity"* for •*"Y Velocity"*, the character should now stop falling once they touch the *ground* or edge of the Stage Window.

Step 4: Jumping

With gravity and Y collisions now incorporated, we can implement jumping! Add another custom My Block: •**"Jump?"**. This will handle our jumping in a nice, organized, and discrete way. Add the ① •**"Jump?"** call between the X and Y change/collision code blocks in the •**[Forever]** loop.

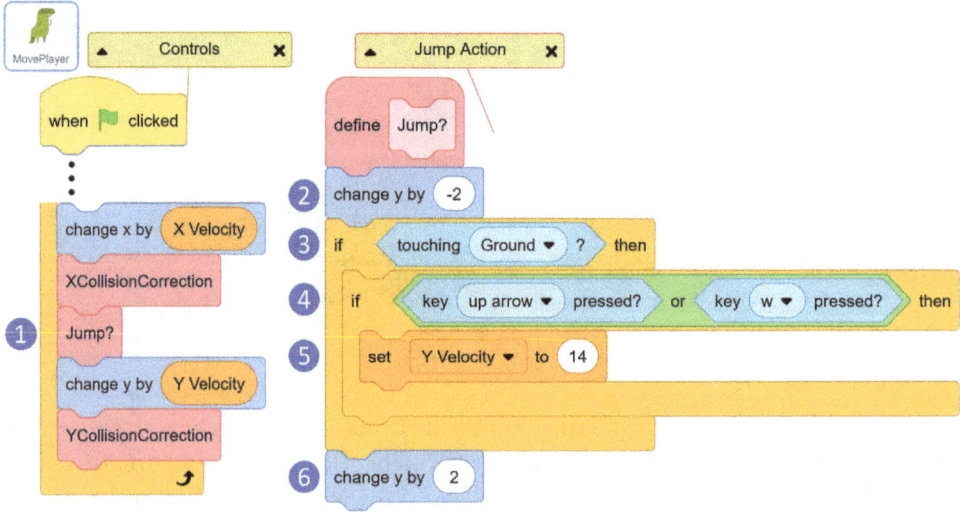

Next, add in the define •**"Jump?"** code. Here we start with a ② •**[Change Y by (-2)]**. This isn't because we want our character to duck, but there's a need to test if they're standing on the ground, so that they have something to jump off of. Without Y-2, they won't trigger a collision with the *ground*

if tested because they are technically above it; they're adjacent rather than overlapping. You could use Y-1, but I use -2 to have a bit more wiggle room. Having Y-2 first, we can then check ③ •**[If** •**<Touching (***Ground***)?> Then]** accurately. If the character is on the ground, we test if the player has given a command to jump with a ④ •**[If** •**<** •**<Key [***Up Arrow***] Pressed?> or** •**<Key [***W***] Pressed?>> Then]**. Again, we'll give the player the option of either WASD or arrow key controls. If they have pushed a jump command button, then ⑤ •**[Set ["***Y Velocity***"] to (14)]**. This gives them enough energy to leap up into the air for a while before gravity pulls them back down. Keep in mind that each frame they move the current •*"Y Velocity"* also each reduces the •*"Y Velocity"* by 1. They're actually moving more of a factorial of the value here, so they end up jumping 105 pixels instead of 14 (each frame they jump one less, so they go a total of: 14+13+12 + . . .). You can experiment with different numbers for different jump heights, but you won't want to increase it much from this to have room for different levels of platforms. Importantly, we also want to undo our initial Y-2, so we ⑥ •**[Change Y by (2)]** to return our character to where they began without anyone noticing.

> **Collision**
> *We don't want our player to be set into the ground, but we have to do that in order to trigger a collision to test if they're indeed on the ground. This kind of technique is more common than you'd think. While in this case we're testing if there's ground under our player, we can do these kinds of tests for other things. Moving something temporarily to test it, get the result, then move back and decide what to do. This hypothetical testing can be very useful in programming to check different options for AI. You can move and test if a collision happens, then move back and decide to move or not. This could check for obstacles for movement, for collisions with enemies or rewards, test for winning moves or any other kind of situational information you need. As long as you can figure out how to determine where to move to and how to move back, you're good. You may need to put these tests into a My Block with no screen refresh to test them rapidly enough (and without other objects reacting to the test move), but it's a powerful technique to know about. We'll explore screen refresh a few steps further in the project so you can get a working model of that.*

Step 5: Drawing the Player

For this project, we're actually going to use two separate sprites to be the player. *MovePlayer* will handle the movement, control input, and collisions, while a new **Dinosaur4** sprite will be renamed *DrawPlayer* and handle displaying the graphics. But why can't we just use one sprite? The reason is that how a character is drawn isn't necessarily how you want it to interact with the environment and other objects. You'll notice our dinosaur has a long tail. When we base our collisions on that dinosaur sprite, the full length of the tail collides with things. This means we can jump up and land on a platform with just our tail and it'll actually hold us up! This is just one example of how silly some collisions can be. To avoid this, we can use a technique called a *collision mask*. In this Scratch adaptation of the technique, what we're going to do is have our dinosaur sprite draw wherever the player is, but the collisions are going to be handled by a differently sized and differently shaped sprite hidden underneath our dinosaur. This will keep our pesky tail from catching on things.

First off, create the *DrawPlayer*, if you haven't already. To start, choose a sprite with **Dinosaur4** from the Sprite Library. You may want to customize the images a little, but you'll need to set the dinosaur to stand on the centre reticle in each costume. This will be our graphical display for the player. We'll ❶ •**GreenFlag** to •**[Set Rotation Style [*Left-Right*]]**, •**[Set Size To (35)%]**, set a •**[Forever]** loop, and make sure it's in the ❷ front layer and to •**[Go To [*MovePlayer*]]**. This will ensure it's always wherever the player is and it draws overtop of the collision mask. Set each costume to stand at the level of the centre target reticle, rather than being centred on it.

Then, work on having the *DrawPlayer* display the appropriate **costume** and **direction** based on how the player is moving. This will give our character wonderfully expressive movement. We can test ❸ •**[If •<•("X Velocity") > (0)> Then]** to •**[Point In Direction (90)]**, and another •**[If •<•("X Velocity") < (0)> Then]** to •**[Point In Direction (-90)]**. Don't use •If/Else here because it would push 0 to one direction, which would switch our left-moving dinosaur to face right when they finished moving, or vice versa. Instead, we keep each as its own separate condition, to only change when needed.

The next condition we'll adapt to is if the character is moving or still. ❹ •**[If •<•(round•(•("X Velocity") * (2))=(0)> Then {} Else {}]** is a complicated statement. What it's doing is seeing if the character has very little •**"X Velocity"**. Round as a function will round off decimals, so any •**"X Velocity"** < .5 would = 0 if rounded. In this case, we double •**"X Velocity"** and round it so only an •**"X Velocity"** <0.25 would trigger this •**If**. This ensures that only the slowest states of the player will trigger it. If standing still, we choose

Advanced Project 3: Platformer Game ◆ 77

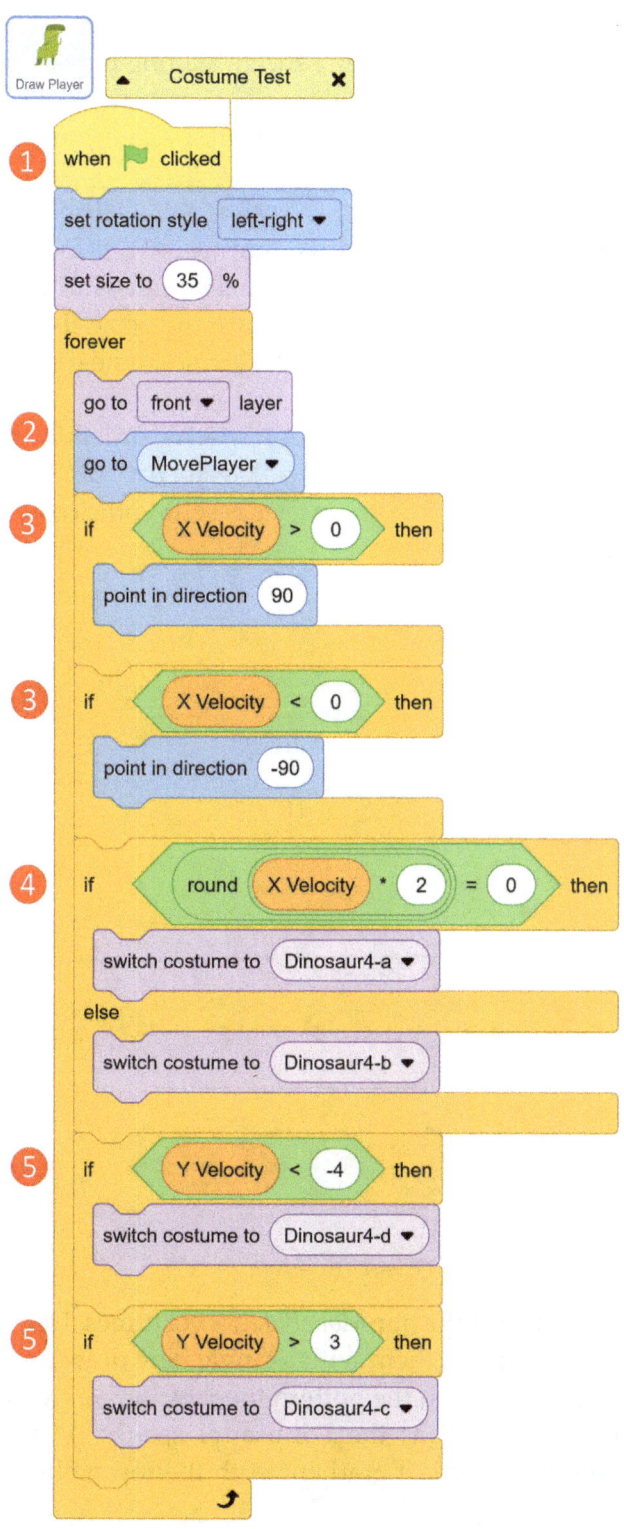

78 ◆ Advanced Project 3: Platformer Game

one costume; if moving faster than a chosen speed, then a different costume displays. This will give our dinosaur a sort of skating movement appearance. You could trigger movement to go through a walk cycle of costumes if you combined what we learned from *Avery's* walk cycle in Project 8: Point-and-Click Adventure.

Add some conditionals that test the ⑤ •*"Y Velocity"* of the player. We can differentiate whether the player is jumping •*"Y Velocity"* >3 or falling •*"Y Velocity"* <-4. You'll see that while the default costumes for **Dinosaur4** are great, we tweaked them a little to better emphasize the jumping. This is good practice for your art skills, and it makes some neat and expressive art for the character.

You'll notice that we did our tests in a specific order. Each test can overwrite the previous condition. So a still •*"X Velocity"* can be overwritten by falling. Since all the output of all these tests is setting the costume of the character, the order is very important.

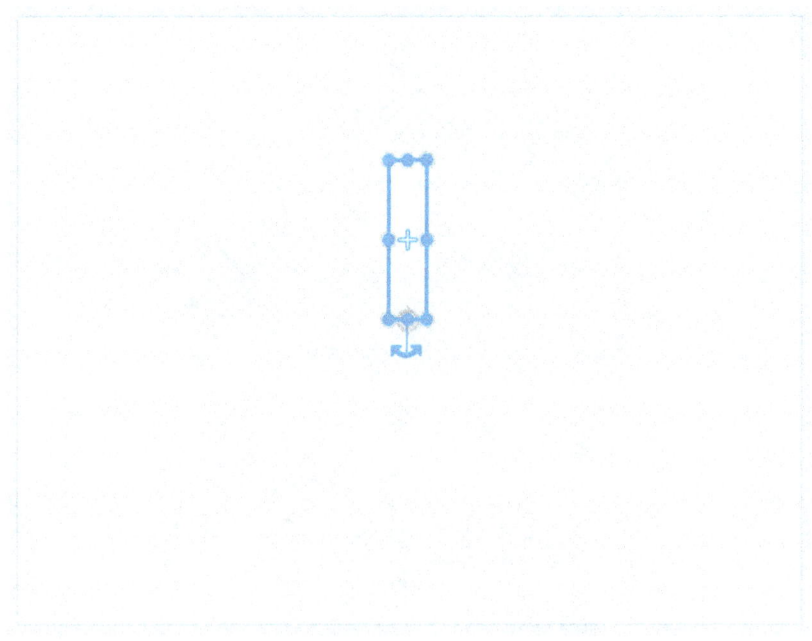

Lastly, we need to change our *MovePlayer's* costumes. Delete all the **Dinosaur4** costumes and replace them with a single new drawing. All we need is a thin tall rectangle placed so the bottom touches the centre target. It should be about two-thirds as tall as the **Dinosaur4** sprites. Make it no outline, and its fill should be a vertical gradient with the top white and the bottom transparent. This should make it impossible to see, but not invisible, which wouldn't allow it to collide in Scratch. The rectangle acts as the blocking or colliding measure for the player. We want it thin and centred so it will collide with walls, but narrow enough that the centre of the dinosaur must be standing on the ground to stay there and not just their tail. Click the 🏁 and try out the new character. Feels and looks pretty good, doesn't it?

If you want an easier way to get started with animation characters or you want to learn more about timing them with sound effects, you can check out the Dino Dance Party project in Book 1: Beginner.

Step 6: Starting Positions

Now that we have our character able to move around the map confidently, we can start making things a little more interesting. First, we'll start by adding our starting positions for our levels. This will relieve us of having to manually reposition our character all the time.

Add a blank sprite "*Start*" and draw in a door, portal, or entry of some kind. I used a greyscale one to make it less showy and distinct. We'll be adding another door/portal for the *exit* so it's easy to just add a more colourful copy of this for the exit later.

80 ◆ Advanced Project 3: Platformer Game

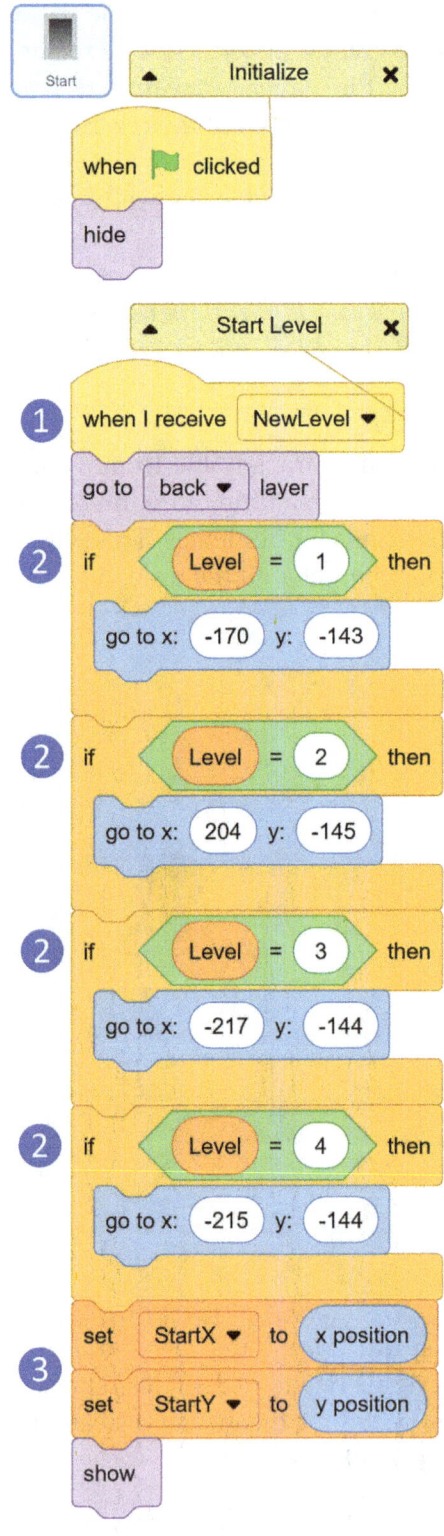

In the code for the object, start by hiding it, then using a ① •*"NewLevel"* event to send it to the back layer (so it doesn't cover up the character). We'll add a new hidden variable •*"Level"* that will track what level the player is on in the game. Then add a series of ② •**[If** •<•(*Level"*) = (#)> **Then]** tests for each level (however many levels you decide to add). In each of those, add a •**[Go To X: (#) Y: (#)]** to set it to an exact location. After the level tests, create two new hidden variables ③ •*"StartX"* and •*"StartY"* and set them to the *Start* door's x/y positions respectively, then •**[Show]** the sprite.

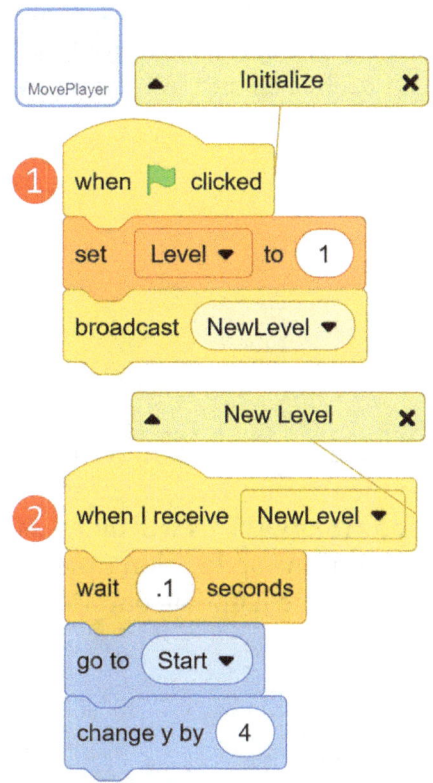

In order for our new •*"NewLevel"* event to work, it needs to be called from somewhere. We'll handle it through the *title* screen eventually, but for now it can be added to our *MovePlayer*. With a ① •**[When ▷ Clicked]** and •**[Set ["***Level"***] To (1)]** and a •**Broadcast ["***NewLevel"***]]**. We want to follow that with a ② •*"NewLevel"* event that contains a •**[Wait (0.1) Seconds]**, •**[Go To [***Start***]]** followed by a •**[Change Y by (4)]**. This will position our character at the start and raise them up so they just have a frame of falling into place as a little accent effect that draws the eye. This will start our game until we handle it properly through the *title* screen system later.

Step 7: Spikes and Damage States

With our start working, we can start getting into trouble. Our next task is adding a threat to the game: spikes! These hazards will impose a challenge to our game by risking a restart for any player that fails to avoid them. You could include any number of hazards in your game, but these will serve as our example for this tutorial. Start by adding a blank sprite, named *spikes*. In its costumes, draw some spikes. You need to position and size each set of spikes in a costume to match with the *ground* in the levels (each level gets its own costume worth of spikes). It's a whole level's worth of spikes in one single costume. Remember, you can reposition selections with either the mouse by dragging or using the arrow keys to carefully step things around the canvas. Create a costume for each level and in the same order as the *ground* costumes so they'll match.

In the code for the spikes, you'll just need a •[Go To X:(#) Y: (#)] and •[Hide] on ① •▷, and then a ② •"NewLevel" event where we'll •[Switch Costume to ["*Level*"] and •[Show]. This will allow us to sync our *spikes* and *ground* to whatever level. But the spikes don't mean much if our character isn't affected by them. So next we'll update our *MovePlayer* to handle that.

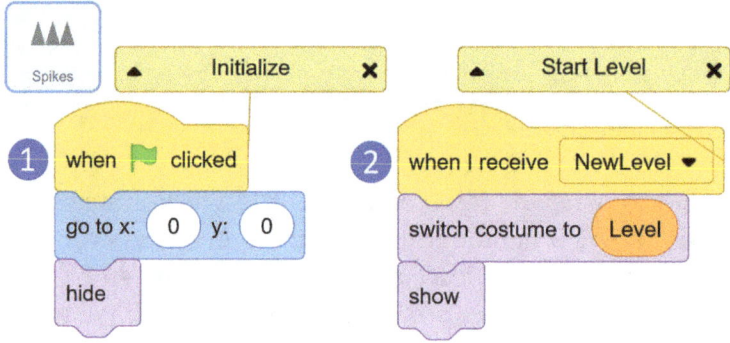

In our *MovePlayer*, add to our •▷ //*Initialize* stack we previously created. Add two new hidden variables •"*Hurt*" and •"*Play*", and at the start we'll add ① •[Set ["*Play*"] to (*True*)]. At the bottom of the stack we'll add a •[Forever] loop. Add a ② •[If •<•("*Play*") = (*True*)> Then] conditional so

Advanced Project 3: Platformer Game ◆ 83

these checks will only occur while the game is actively played, allowing the game to pause these effects if play is false. Add a conditional to test ③ •[If •<Touching [*Spikes*]?> Then] in that we'll •[Set ["*Hurt*"] to (50)] and call a new My Block •"OnDead". We can ④ duplicate this event and place it below the first; instead of testing touching spikes, we'll replace that with •[If •<•(Y Position) < (-179)> Then]. This will test if the character has dropped to the bottom of the screen, falling into one of the bottomless pits that some levels should have.

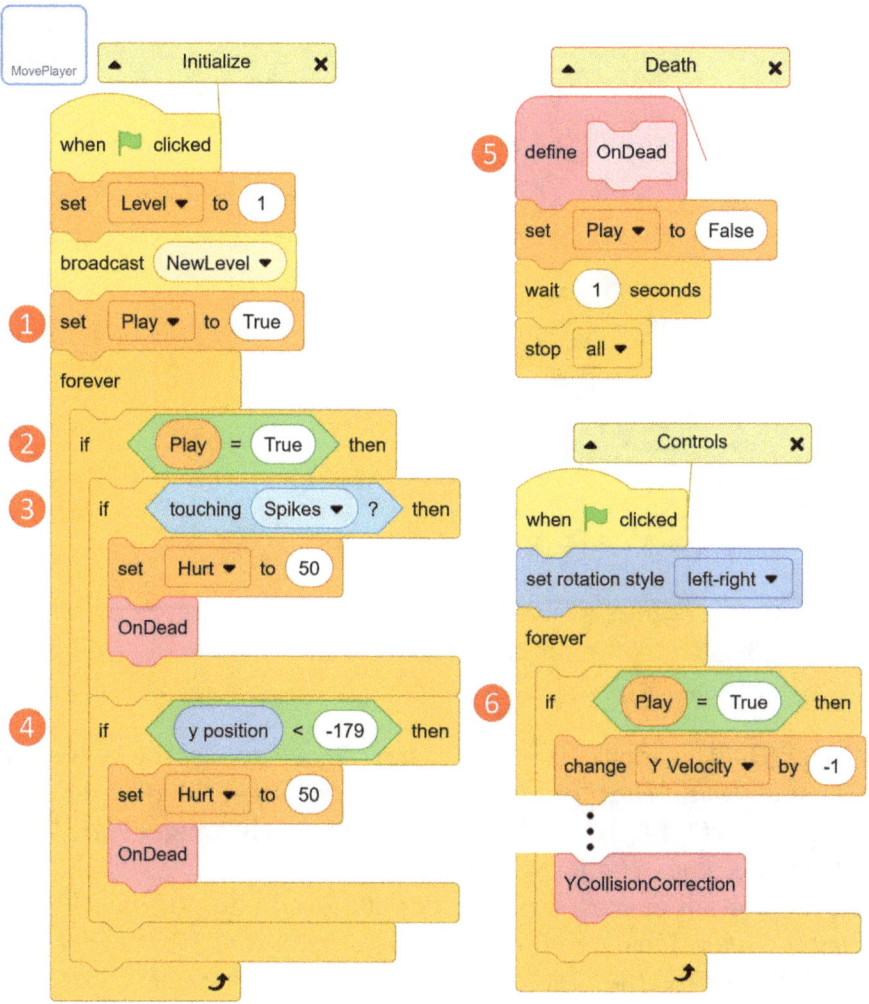

Next, we'll define the custom ⑤ •"OnDead" block that will be able to handle deaths from any of the hazards in the game in a single place. Cancel all movement by setting both velocities to 0 while also •[Set ["*Play*"] to ("*False*")], •[Wait (1) Seconds], and •[Stop [*All*]]. We'll improve on this later, but for now it works. Then, go to the movement //Controls stack and add an ⑥ •[If •<•("Play") = (True)> Then] conditional with everything from •[Change ["Y Velocity"] by (-1)] and down inside it. This way, the character only falls/moves while the game is in play.

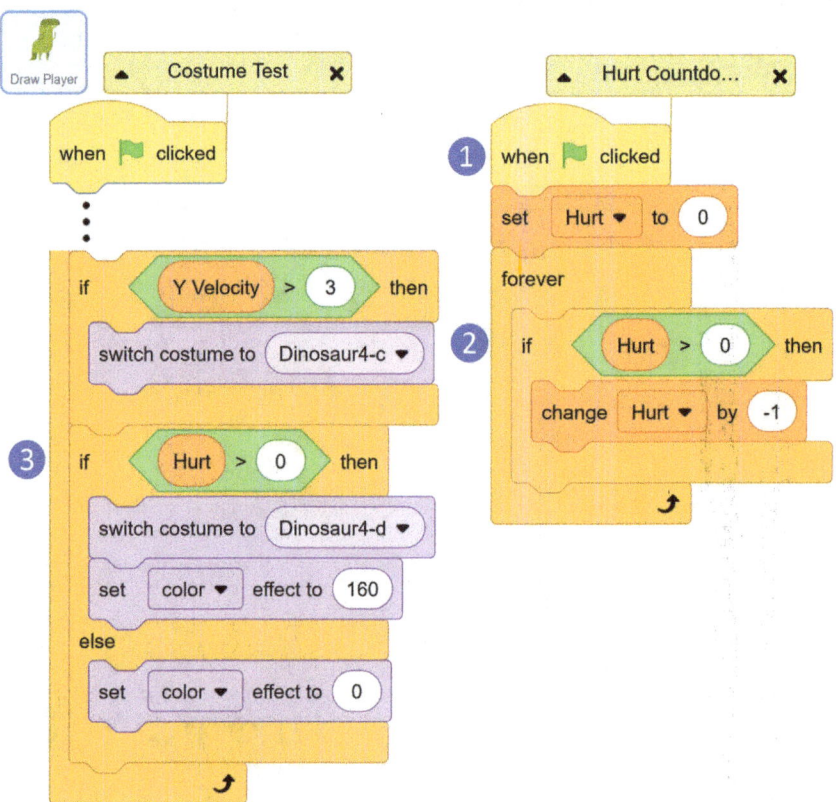

Our •"Hurt" variable is used in our *DrawPlayer* sprite. Here, add a ① ▷ and •[Set ["Hurt"] to (0)]. Then, in a •[Forever] loop, add a count-down for •"Hurt". ② •[If •<•("Hurt") > (0) Then] •[Change "Hurt" by (-1)]. So •"Hurt" becomes a temporary state.

At the bottom of our //Costume Test stack, add an •If/Else, ③ •[If •< •("Hurt") > (0)> Then] with a •[Switch Costume to [Dinosaur4-d]] and •[Set [Colour] Effect to (160)] that will draw our character with its wide open (screaming) mouth and colourize it red. In the •Else {} we just have to •[Set [Colour] Effect to (0)] to reset to normal colouration. With the count-down and these graphic effects, the dinosaur will now shout in pain and then the game ends. We'll make use of the countdown ending the effect after our next step.

Step 8: Player Lives

Now that we've added a deadly hazard to our game, we need to add in a player lives system. This is a common mechanic in games to give players some forgiveness to practice difficult sections without completely losing the game in a single wrong move but still provide the challenge of major consequences. Our player lives system will be managed in the *MovePlayer* sprite.

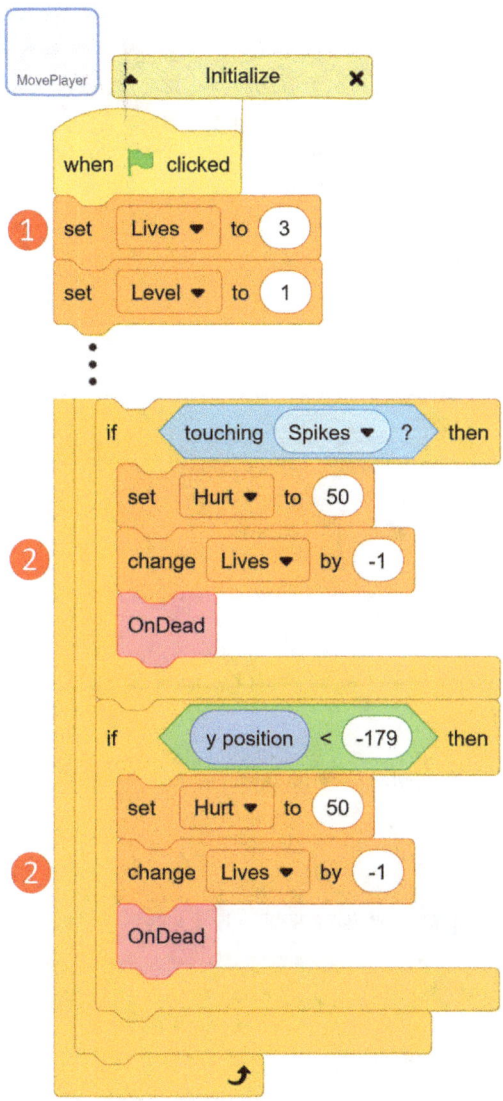

Add in a new visible variable •*"Lives"*, and in our *//Initialize* stack, add a ① •[Set [*"Lives"*] to (3)] at the very top. In ② •[If •<Touching (*Spikes*)?> Then] and ② •[If •<•(Y Position) < (-179)> Then], add a •[Change [*"Lives"*]

by (-1)] so touching spikes or falling to the bottom of the screen will both cost the player a life.

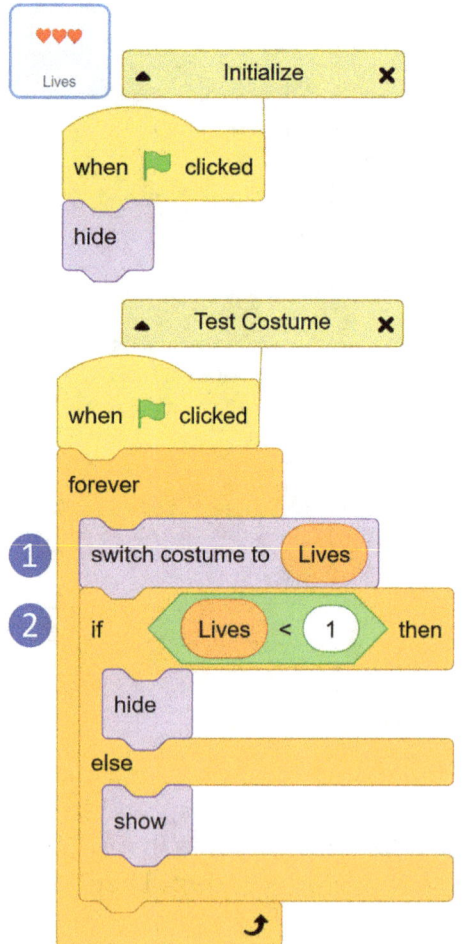

We can add a new blank sprite called *"Lives"* to act as a graphical display of how many lives the player has. You could just use the lives variable, but a graphical display is a nice touch and good to know how to do. In the costumes for this sprite, make the first costume one heart, the second two hearts, and the third three hearts. The code sets the ① costume to the number of lives. Because the costumes start counting from 1, ② if the player has 0 lives, we simply hide the sprite instead since we can't have a costume 0. With this working, you can make the •*"Lives"* variable hidden.

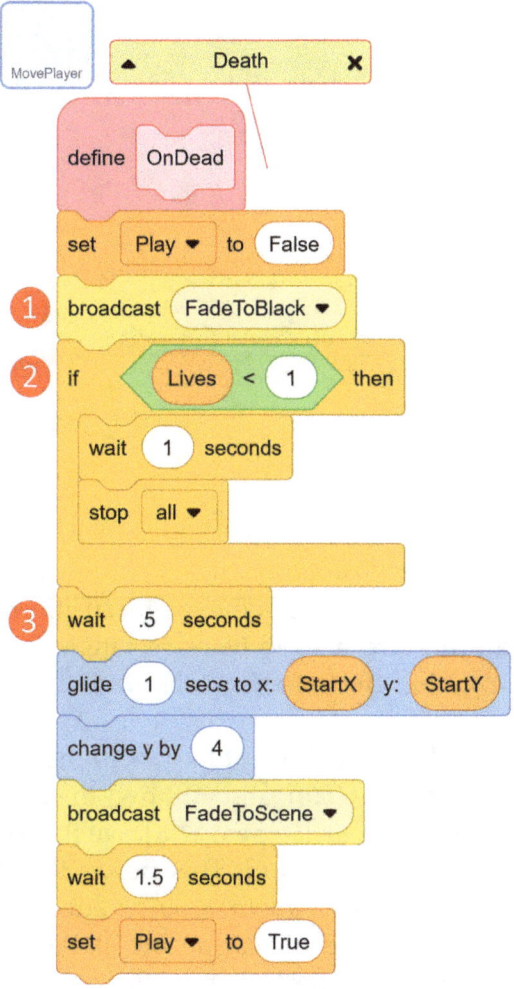

Back in *MovePlayer*, in •[**Define** *"OnDead"*], below •[**Set** [*"Play"*] **to** (*False*)] add a ① •[**Broadcast** [*"FadeToBlack"*]]. Then, add an ② •[**If** • <•(*"Lives"*) < (*1*)> **Then**] with the •[**Wait** (*0.1*) **Seconds**] and •[**Stop** [*All*]] inside. This will end the game when you run out of lives, but it adds the wait in to make sure the character changes to their •*"Hurt"* state for dramatic effect. Below the •**If**, ❸ add a •**[Wait 0.5 Seconds]** and add in a •**[Glide (1) Secs to X: ("StartX") Y: ("StartY")]**, •**[Change Y by (4)]** and •**[Broadcast ["FadetoScene"]]**. Then, •**[Wait (1.5) Seconds]** and •**[Set ["Play"] to (True)]**. Our code below the •**If** allows the game to resume by moving the character to the start position and, after some delays, turning the •*"Play"* variable back to *"True"*, unlocking the controls. If they have 0 lives, though, the game stops.

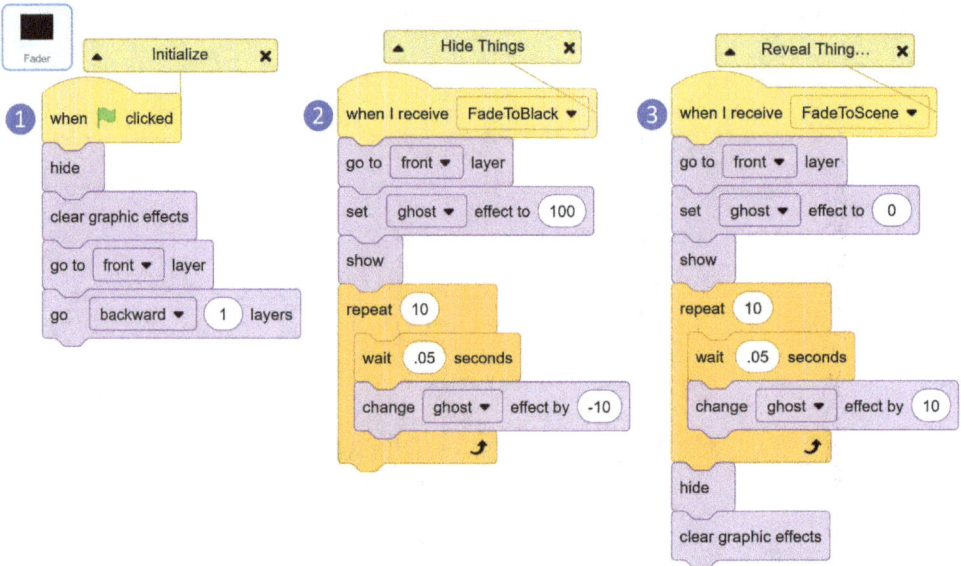

For the two •**Fade** events, adding a new blank sprite *"Fader"* positioned to X: (0), Y: (0). We'll use this to do some fade transitions, so you can make its costumes just a solid screen-sized black rectangle. For its code, ❶ •⚑ **[Hide]**, •**[Clear Graphic Effects]**, •**[Go to [Front] Layer]**, and then •**[Go [Back] (1) Layer]**.

It needs a ❷ •*"FadeToBlack"* event to darken the screen. Send it to the front, •**[Set [Ghost] Effect to (100)]**, and •**[Show]** to start with a completely transparent version. Then, in a •**[Repeat (10)]**, we'll •**[Wait (0.05) seconds]** and •**[Change [Ghost] Effect to (-10)]**. This makes the sprite become fully opaque in ten frames. The ❸ •*"FadetoScene"* event reverses this. Have the •**Ghost Effect** starting at 0 and increasing by 10. After the •**[Repeat]**, we •**[Hide]** the sprite and then •**[Clear Graphic Effects]**.

Both events make the fader go to the front layer, but only when the event is first called, whereas the *DrawPlayer* goes to front every step, which will put the fader behind the character but in front of everything else, which will provide a great dramatic highlight of the player.

Step 9: Waypoints

With player lives now incorporated, let's add in some amount of progress saving in the game. This system will double up in value by also working as a key system. Players starting a level will only be able to see the *waypoint*, which acts as a progress-saving beacon, but touching it will also reveal where the *exit* to complete the level is. This way, there's some hidden information to the level so players can't strategize too far ahead, adding more of an exploration feel to the game.

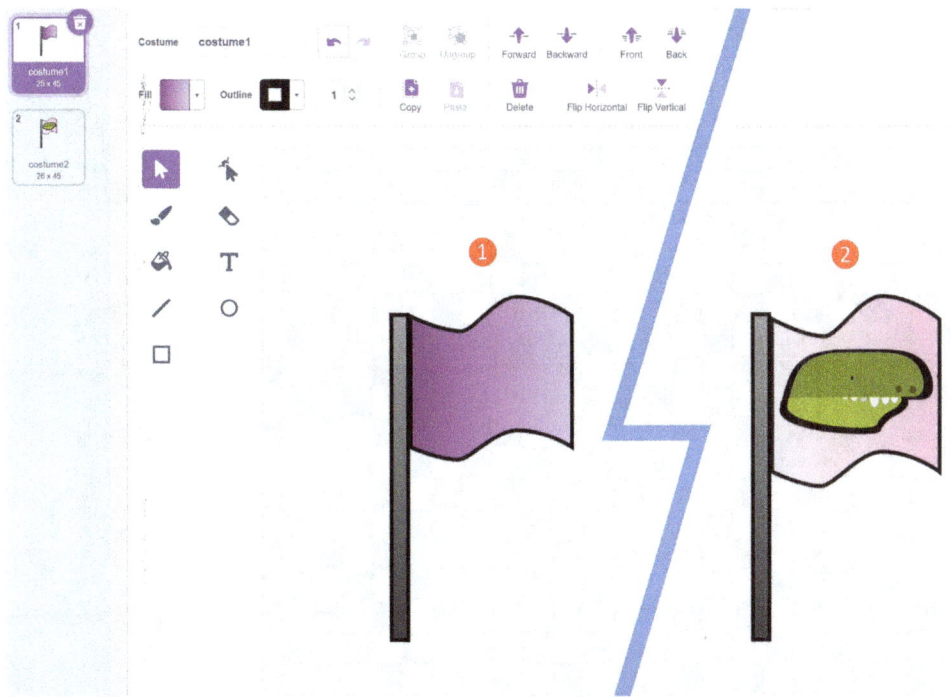

Add a blank sprite, *waypoint*, and make two costumes for it. The ① first will be a dull or dark flag; the ② second, a brighter success flag. Or you could even just do one flag and have the first just be the bare flagpole. Whatever your design, it should make it obvious to the player that they succeeded/achieved something when it changes after they touched it.

In the code, add a start hidden combo, then we'll need a ① ●🏳+●**[Forever]** loop, testing if the *MovePlayer* has touched it. When they have, it'll switch to the touched costume and give a congratulatory message to the player. Also, change ② ●*"StartX"* and ●*"StartY"* variables to the ●**X Position** and ●**Y Position**, respectively, of the *waypoint*. Remember, these are the coordinates that a player returns to when they lose a life. This way, after reaching

90 ◆ Advanced Project 3: Platformer Game

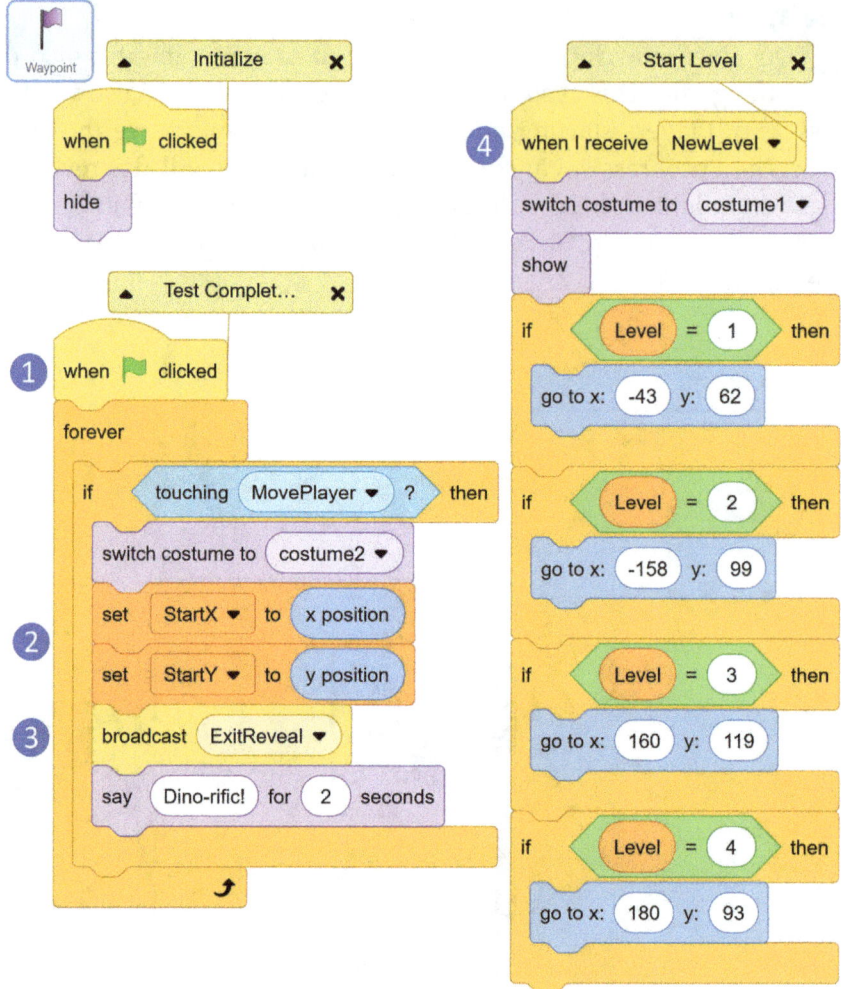

a waypoint and losing a life, they'll return here instead of the *Start* door. End by ③ •[**Broadcast** [*"ExitReveal"*]], which we'll make use of later.

In a ④ •*"NewLevel"* event, have the *waypoint* switch to its untouched costume and •[**Show**], then use a series of •[**If** •<•(*"Level"*) = (#)> **Then**] conditionals to determine what level is being loaded and have •[**Go To X:(#) Y: (#)**] blocks in each to position the *waypoint* for that level.

> *A great effect for this achievement would be using fireworks! You can check out how you might want to achieve that in our Fireworks project in Book 1: Beginner!*

Step 10: Exit and New Level

Now that we've gotten to the *waypoint*, our next goal is the *exit*. Let's duplicate our *start* sprite and rename it *"exit"*. In the costumes, recolour the exit to be more interesting-looking than the *start*.

In the code, delete the •[Set] •"*StartX*" & •"*StartY*" blocks. We'll need a ① •▷+•[Hide]+•[Forever] loop. In it, test ② •[If •<•<Touching (*Move-Player*)?> Then] and •[If •<•<Key (*Up Arrow*) Pressed?> Or •<Key (*W*) Pressed?>> Then]. We could combine these in a single **If** statement, but we aren't, for reasons you'll see later in the project, so for now just nest one inside the other. These conditionals will mean the player will need to reach the exit and press Up/W to go through it. Use a ③ •[Change ["*Level*"] by (*1*)], •[Broadcast ["*NewLevel*"]] and •[Hide]. This will trigger all the objects' •"*NewLevel*" codes that set them up for a new level, but make sure to increase the level beforehand so it isn't just the same level that is reset. Hide the *exit* because the player has to earn the reveal via the next level's *waypoint*. Adjust the ④ •"*NewLevel*" event for the *exit* to position the door correctly per level, and ⑤ have any level number above our planned levels broadcast a •"*Win!*" event.

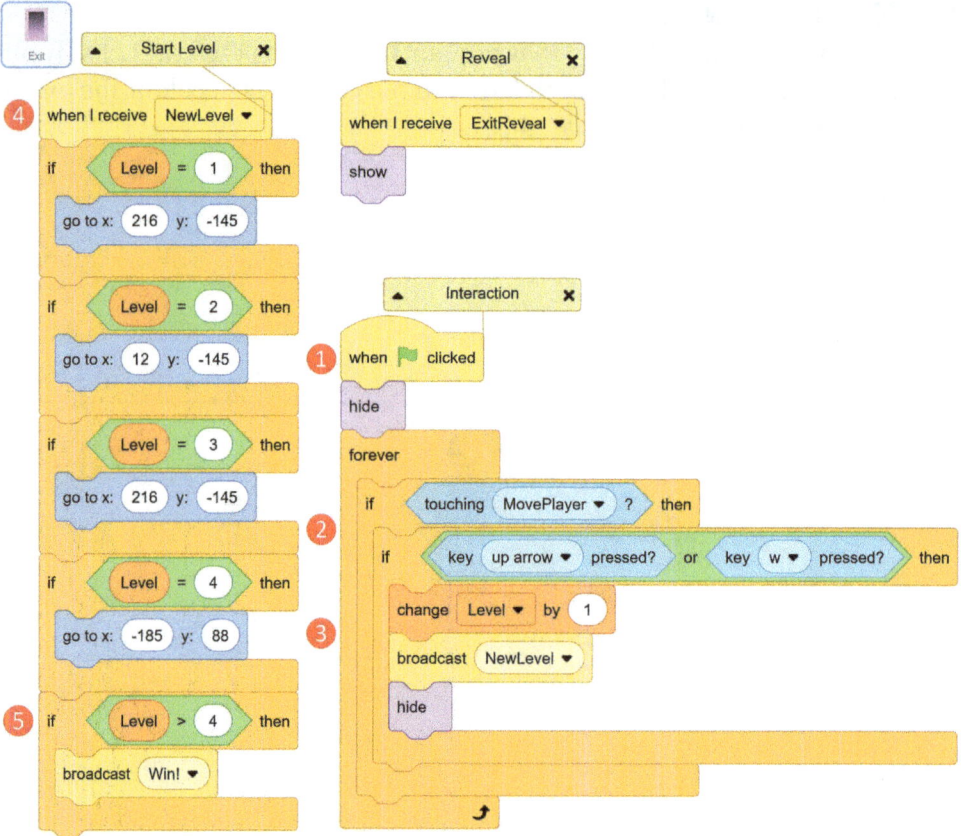

Now, to make sure everything is loading on •*"NewLevel"* event! In our *ground* sprite, add some code now that we're switching levels. In a ① •[**When I Receive [***"New Level"***]**] event, •[**Switch Costume to [***"Level"***]**], again using the •*"Level"* variable as the number of the costume. Next, run our •[**Go To [***Back***] Layer**] code block to ensure it's still always at the back. As long as we've added the right position data and costumes, our game can adapt to any level and go for as long as we want.

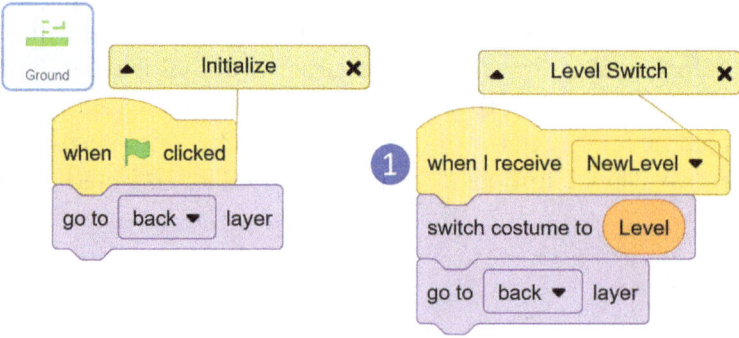

Step 11: Bouncer Jump Boosting Platforms

The reason platformers are so fun are all the different obstacles, hazards, and designs one can create, find, and explore. A big part of this is making interesting obstacles or aids that can offer different opportunities to tackle the terrain or avoid hazards. Now that we've covered the basics, let's add a bunch of these to provide a tool set for creating interesting levels, starting with a bouncer.

Add a blank sprite, *bouncer*. This sprite will provide extra jumping power to the character when they stand on it. We'll use the same technique we did for the ground. Create a costume for each level and draw in areas where you want the player to be able to jump extra high. In the code, we'll start it ① hidden, and we'll need a ② •*"NewLevel"* event with a •**[Switch Costume to [***"Level"***]]** and •**[Show]**. This will have the bouncer display on each level with the appropriate costume. Next, edit the *MovePlayer* so they can make use of it.

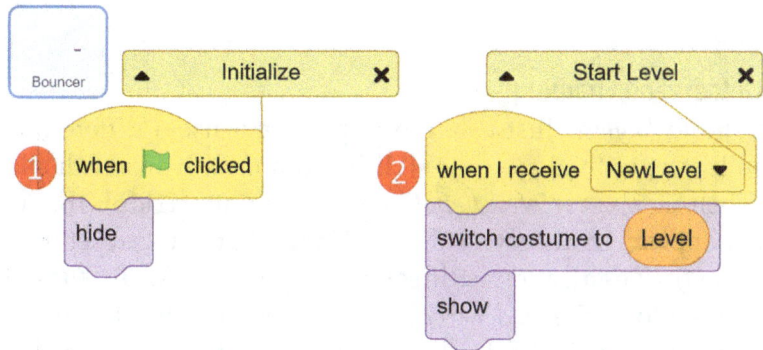

In the *MovePlayer* **Code** tab, go to the *//Jump Test* stack. Change up the code inside the •**[If** •**<Key (***Up Arrow***) Pressed?> Then]** conditional. Add an ① •**If/Else** of •**[If** •**<Touching (***Bouncer***)?> Then {**•**[Set** *"Y Velocity"* **to (20)]} Else {**•**[Set [***"Y Velocity"***] to (14)]}**. Now our normal jump of 14 is boosted to 20 if the player starts their jump while touching the bouncer.

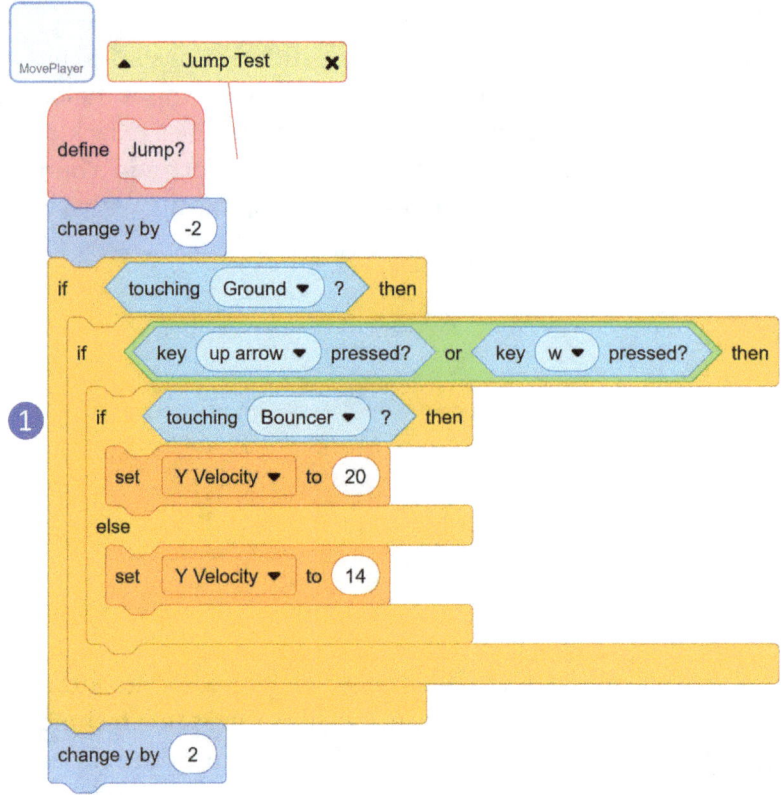

When designing your levels, keep in mind that bouncers don't count as *ground*. You can't stand still on them as they do not stop you from falling, so you'll need to position them overtop the *ground* for the player to interact with both.

Step 12: Ladders and Climbing

Next, we'll add *ladders*. Just as an example, we'll use a different technique for it. You'll start with a ① •▷+•[Hide] combo so it won't display unless needed. Then in a ② •"*NewLevel*" event, have it •[Hide]; this will make sure it only appears in the correct levels. Under that you can set up an •[If •< •("*Level*") = (1)> Then], and inside set it to a specific •[Go To X:(#) Y:(#)] then •[Show]. This allows the *ladder* to be a single costume, but it can only be in a single position per level. This isn't as versatile, but it provides an example of a very minimal system. Feel free to use the *bouncer's* code instead and have full level designs for multiple ladders with level-specific costumes if you prefer, but I wanted to show you this lightweight alternative. The other advantage of using this method is that you could use costumes for different shapes or sizes.

Advanced Project 3: Platformer Game ◆ 95

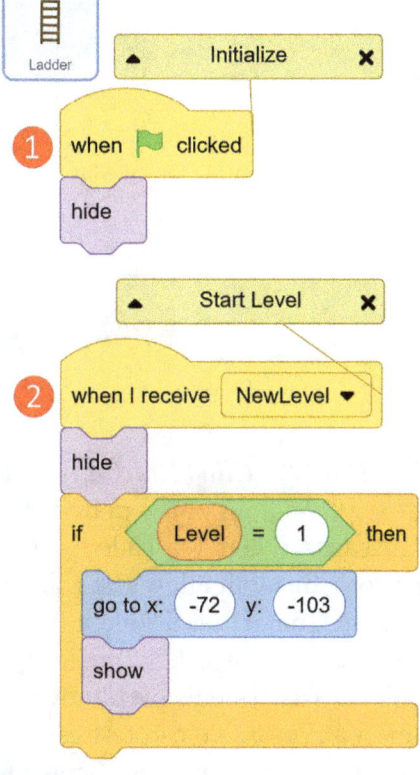

To use ladders, in the *MovePlayer* sprite, we will need to alter the •[Define *"YCollisionCorrection"*] stack. Beneath the •[If •<Touching [*Ground*]?> Then] conditional, add a new ① •[If •<Touching [*Ladder*]?> Then] conditional. Inside it, we need an •[If •<•<Key (*Up Arrow*) Pressed?> or •<Key (*W*) Pressed?>> Then]. Then, inside that one add a •[Change Y by (*4*)]. This will allow the player to climb up the *ladder*. We'll duplicate this and switch the keys to the down arrow and S key. In the duplicate, ditch the •[Change Y By (#)] code block and use a custom My Block ② •*"ClimbDown"* – when creating this custom block, add an input (which will create a value or parameter for how fast something climbs when using this function), as well as make sure to check the **"Run without Screen Refresh"** checkbox.

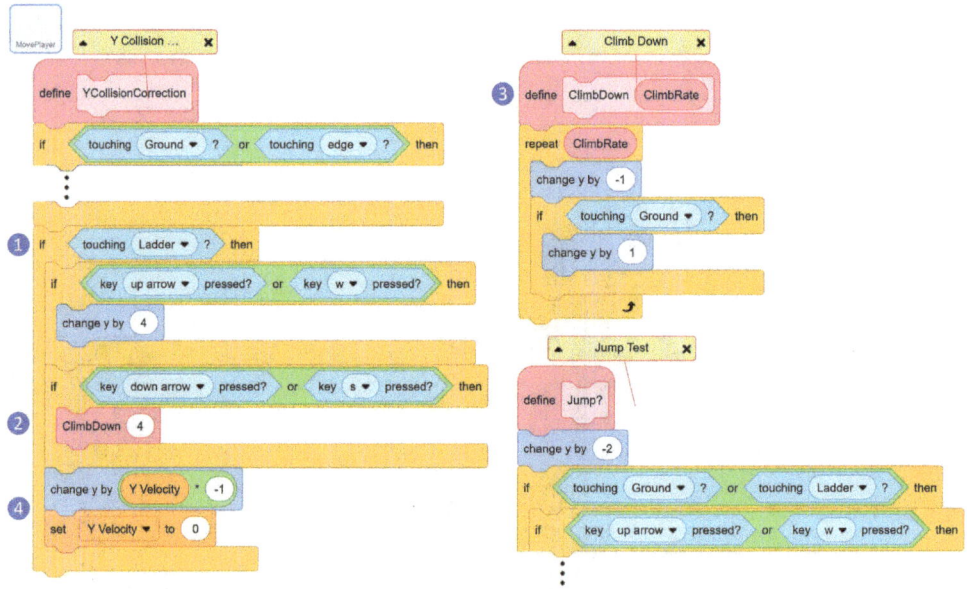

In ③ •[Define *"ClimbDown"* (*"ClimbRate"*)], do a •[Repeat (*"ClimbRate"*)] (just pull a copy of (*"ClimbRate"*) out of the define block to put in the repeat) •[Change Y By (*-1*)]. Because when we called •*"ClimbDown"*, we set its *"ClimbRate"* parameter to 4, this will allow the character to climb down at a rate of 4 pixels per frame. But what if they are already at the bottom? We need to test if they've moved too far. Add in a •[If •<Touching (*Ground*)?> Then] with a •[Change Y By (*1*)]. This runs each single pixel moved down and tests if it was too far. By putting them in a repeat, we can test each single pixel moved, but because the •My Block runs without waiting for **Screen Refresh**, all four repeats are run in the same frame with no delay. This way, we can move as quickly as we want but still be pixel-perfect with our collision detection. This screen refresh skipping is the reason we need the code inside a custom block instead of in the main stack.

> **Screen Refresh**
>
> *The term "screen refresh" here refers to the act of drawing the game on your monitor. Scratch runs code every frame, or 1/30th of a second. When the code runs, various data can change the properties of sprites, variables, etc. After all the code is run for that frame, Scratch then draws the results. This is the screen refresh. The last frame was drawn and kept on-screen until the next frame resolves and all the properties and variables are updated, then Scratch overwrites the previous view of the game world or screen-refreshes. This gives the latest view of the game world as it stands. If code runs without screen refresh, some code blocks like repeat are run to completion without waiting one frame of animation between each iteration. This means we can get our handy animation use of repeat to do only one thing per frame by default, but if we put it into a My Block that skips screen refresh, we can have it blaze through its tasks at top speed and resolve everything before it does a screen refresh. To get the best of both worlds, we just needed to know how to access this alternative, and very powerful, option. Just make sure you don't use any forced delay blocks (say for x seconds, wait x seconds) within a "Run without Screen Refresh" My Block or you'll defeat the purpose.*

In the **[If <Touching (*Ladder*)?>]** test below the **[ClimbDown] If**, we need a ④ **[Change Y by (("*Y Velocity*") * (-1))]** and a **[Set ["*Y Velocity*"] to (0)]**. These will stop the player from falling while in contact with a *ladder*, essentially making it act like *ground*. If we didn't have these here, our player would simply continue to fall down the ladder when standing on it, or even if they were trying to climb up it.

We also need to update our **[Define "*Jump?*"]**. As we want our *ladder* to function like ground, we should be able to jump off it. In the **[If <Touching [*Ground*]?> Then]**, add a **<<condition> Or <condition>>** block, also including **<Touching [*Ladder*]?>** to it.

Step 13: Moving Platforms

One of the classic obstacles in a platformer is the moving platform. These are also famously difficult to program. We've come a long way, so let's try tackling the challenge! Again, we're going to try another method for creating the different level versions of the object here. This time we're going to use cloning so that we can have multiple independent objects operating on a single level. Start by creating a blank sprite, *MovingPlatform*, and for its only costume make a suitably sized rectangle of some distinct colour.

We're going to need two •*"NewLevel"* events. One will just be a ① •**[Delete This Clone]** stack. The other will be used to generate clones by level but importantly starts with a ② •**[Wait (.01) Seconds]** code block. Each level will need to destroy the previous level's clones, so by putting the minimal delay in the second event, we ensure the deletion happens first, then we can create the new platforms for the current level.

Use the same ③ •**[If** •<•**(***"Level"***) = (#)> Then]** conditionals for each level we need moving platforms in. Within each, use •**[Go To X:(#) Y: (#)]** to move to where we want a platform to start and then ④ •**[Create Clone of [***Myself***]]** to generate a new clone. Then, set two new hidden variables ⑤ •*"PushX"* and •*"PushY"*. These variables control the movement of the *moving platforms* in the level. Unfortunately, all the *moving platforms* will need to move the same, but we can have them move horizontally, vertically, or diagonally by setting the •*"PushX"* and •*"PushY"* to some combination of 1, 0, and -1. You'll be able to experiment with different combinations to see how they work.

In the ⑥ •**[When I Start As A Clone]** event, we'll need a •**[Show]** (the original always stays hidden) and a •**[Forever]** loop. Inside it, add a conditional to test •**[If** •<•**(***"Play"***) = (***True***)> Then]** to ensure the platform only moves while the game is in play. Inside that conditional, ⑦ •**[Change]** X and Y by the •*"PushX"* and •*"PushY"* variables with a •**[Repeat (***100***)]** loop. This will make the platforms move 100 pixels in the appropriate direction from where they started. Then, to ensure they move back to where they began, ⑧ •**[Set [***"PushX"***] to** •**(**•**(***"PushX"***) * (-1))]** and the same for Y. This *-1 operation

Advanced Project 3: Platformer Game ◆ 99

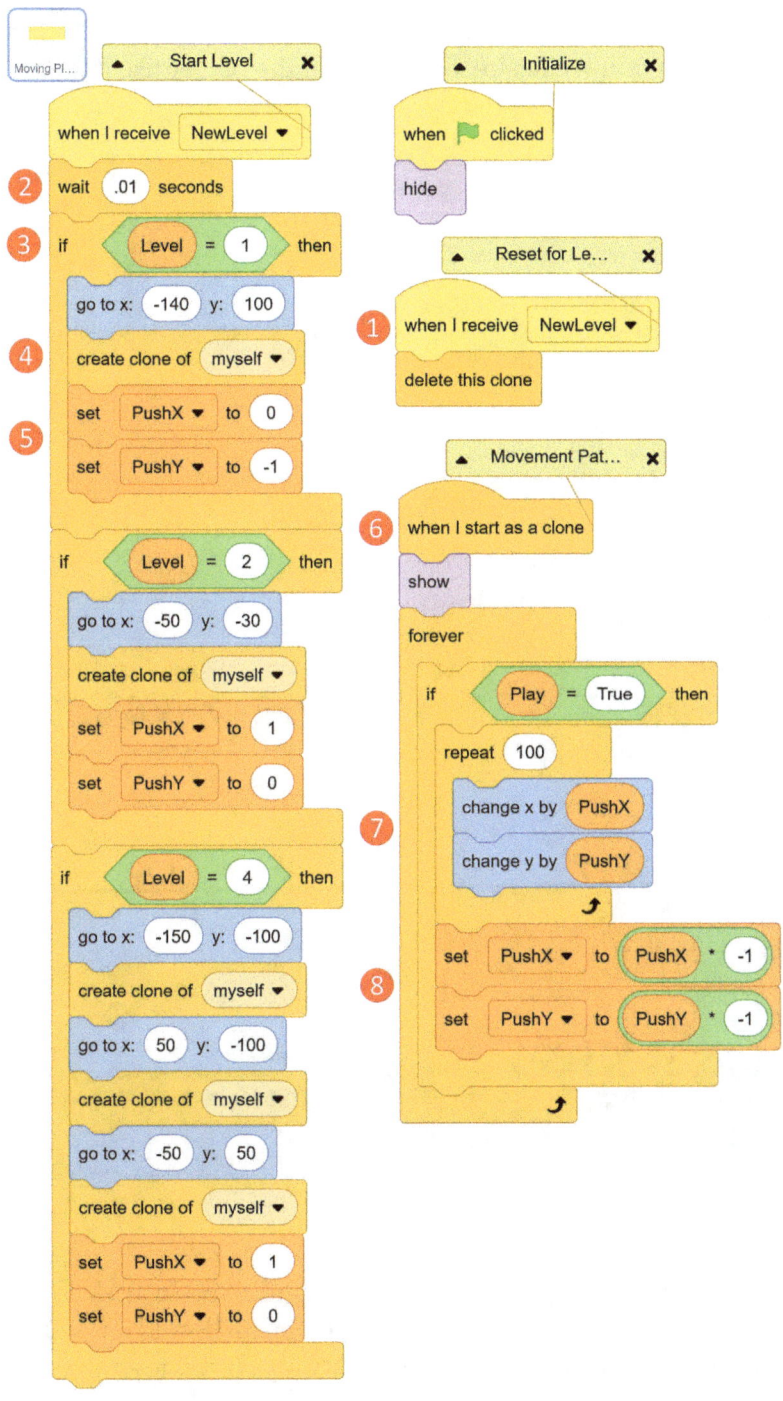

will reverse a direction, a 1 becomes -1, and vice versa. The next run through the ●[**Forever**], the movement directions will be reversed for the ●[**Repeat (*100*)**]. Our *moving platform* now moves back and forth for eternity in any level in whatever direction we desire, but how can the player use it?

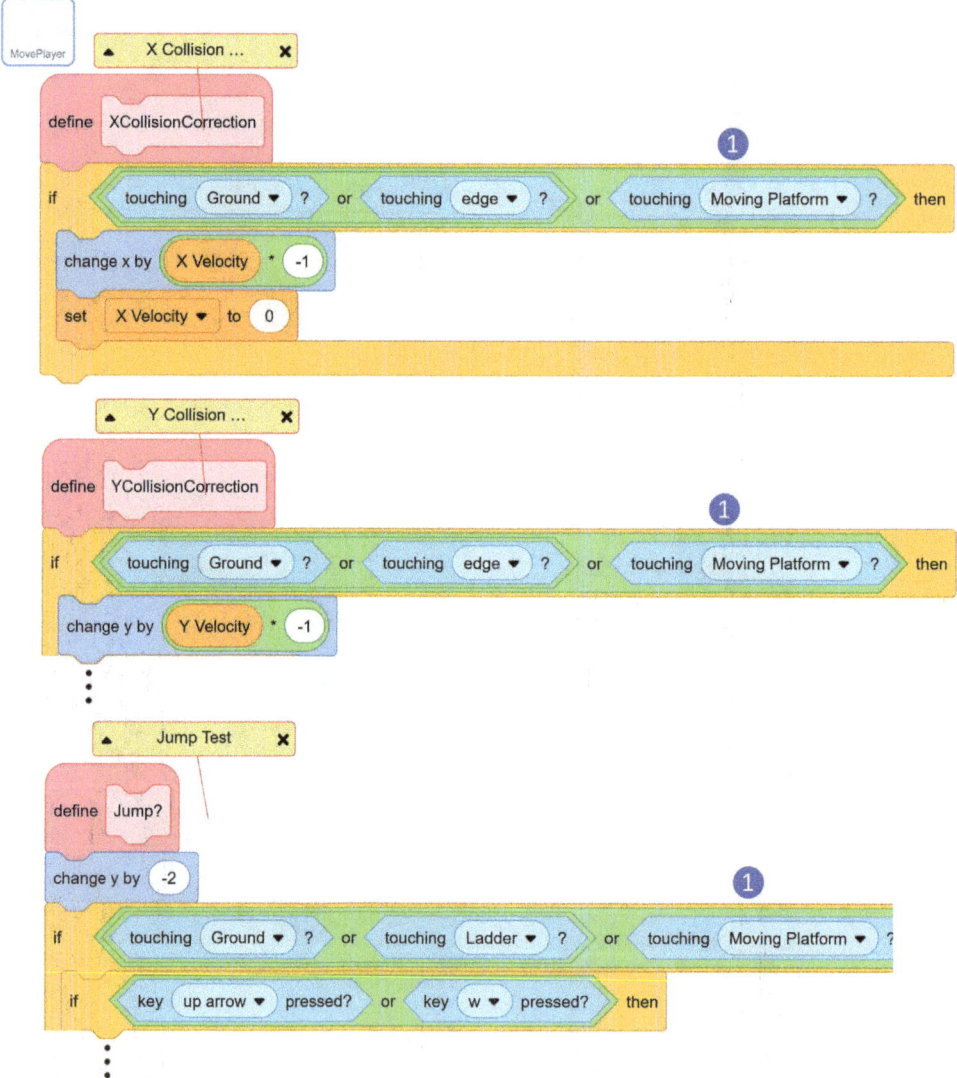

First off, we'll need to get the player able to land on it. Go to *MovePlayer*; in ❶ every stack where you test with ●<**Touching (*Ground*)?**>, add an ●<**Or** ●<**Touching (*Moving Platform*)?**>>. This will allow *moving platforms* to act as solid ground.

But if you notice, the player eventually falls through the platform when it moves up, because it doesn't push the player up (or left/right if you test one that moves sideways). We need to add a couple of things in order to make that happen.

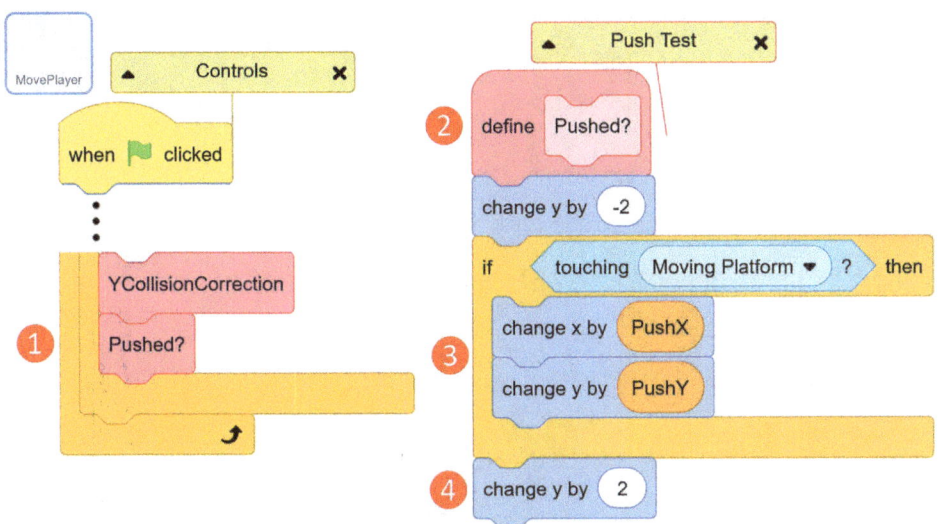

In the movement *//Controls* stack, add a new custom My Block ① •"Pushed?" under the •"YCollisionCorrection" call. In ② •[Define "Pushed?"], add •[Change Y By (-2)]. This is the same technique we used earlier to detect what the player is standing on. An •[If •<Touching (*Moving Platform*)?> Then] will make this run only with a *moving platform* collision. Inside it, ③ •[Change] X and Y by •"*PushX*" and •"*PushY*", respectively. This will let the direction variables of the *moving platform* affect the player and move them at the same rate and in the same direction of the *moving platform*. Beneath the •If, add a ④ •[Change Y by (2)] to reset the player to their true, original position.

Step 14: Blinking Platforms

The last obstacle we'll add to our game are blinking platforms, or platforms that appear and disappear like clockwork. These platforms force the player to time out their movement; generally used to jump over dangerous gaps, they add tension to the game with a degree of time-sensitivity.

Add a new blank sprite, *blinking platform*, and for its costume make level-specific costumes just like *ground*. At each position that we want a blinking platform, add in an appropriately sized rectangle of some distinct colour.

102 ◆ Advanced Project 3: Platformer Game

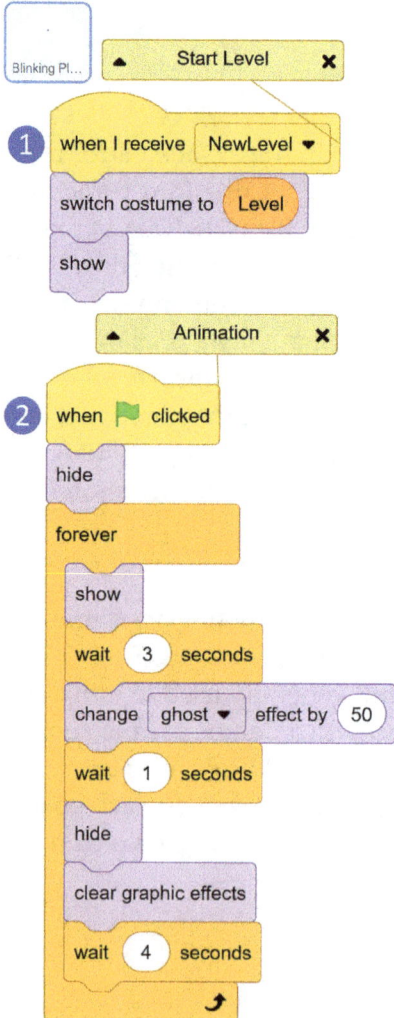

In the code, add a ① •*"NewLevel"* event to •[**Switch Costume to** •(*"Level"*)] and •[**Show**]. In the ② •▷+•[**Forever**] combo, add a •[**Show**], •[**Wait** (*3*) **Seconds**], •[**Change** [*Ghost*] **Effect By** (*50*)], •[**Wait** (*1*) **Seconds**], •[**Hide**], •[**Clear Graphics Effects**], and •[**Wait** (*4*) **Seconds**]. This creates a three-phase cycle for our platform. It shows for three seconds, then is only half opacity for 1 second to create a warning that it will disappear shortly, then it disappears, and after 4 seconds, it repeats the cycle by reappearing. You can play around with the timing of this cycle to adjust difficulty.

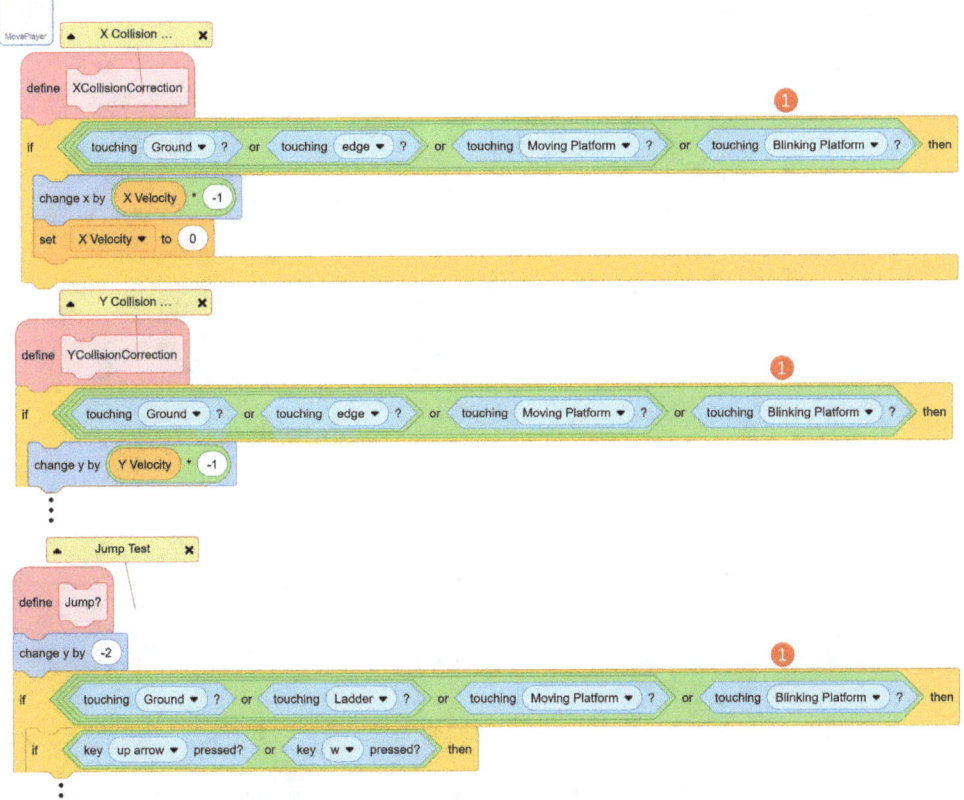

Just like the *moving platform*, we have to adjust the *MovePlayer* sprite's code ① to react properly to the *blinking platform*. Anywhere you can see •<**Touching** (*Ground*)?>, add an •<**Or** •<**Touching** [*Blinking Platform*]?>>. Now the character can stand on the *blinking platforms*, but only when they're visible. A sprite that is hiding doesn't cause or receive any collisions, so when it hides, the player will fall through the now-empty space if they haven't jumped away. If the long chains of different conditions aren't to your liking, you could also duplicate the Ifs and have one check for two types of ground and the other check the others; one will stretch your code horizontally, the other vertically.

Step 15: Victory!

So what happens when we've run through all our levels? We better add a *Win* or *Victory* screen and end the game. You'll notice our *exit* sprite's code is built to handle when the player has run out of set levels. At the end of the ●*"New-Level"* stack, it has a conditional ●[If ●<●("Level") > (#)> Then] that tests for any level number beyond our last included level. If your original code for the *exit* is still valid, you're good, but if you've added more levels, you'll need to update this code by setting the correct number for the maximum level. This will run once the player has finished all the set levels, and it increments past the last level calling the ●*"Win!"* event.

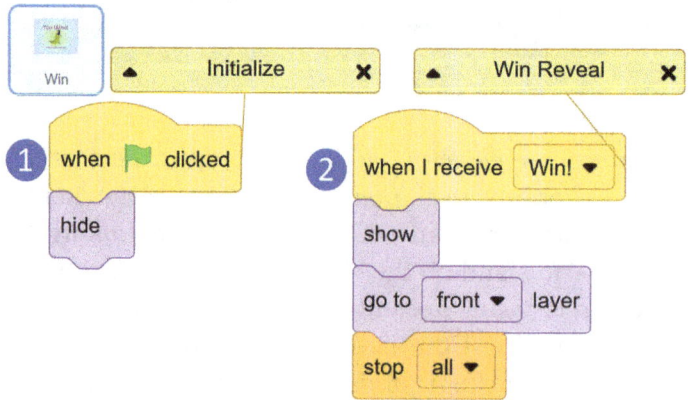

To handle this event, add a new blank sprite, **"Win"**; as usual with new sprites, move it to X: 0, Y: 0. We aren't going to move it, so just setting its properties is fine. We can make its costume be our *Win* screen, congratulating the player and thanking them for playing, but you could also add in some expository to resolve a story too. In the **Code**, ❶ start with the sprite hidden. Then, on ❷ •**"Win!"**, have it •**[Show]**, •**[Go To [*Front*] Layer]**, and •**[Stop [All]]**. Simple but effective.

Step 16: Title
Lastly, add a *title* screen to our game. We don't have any settings to change, so it's not a full main menu, just a title card to introduce it. Add a blank sprite, *title*, and in the costume create a *title* graphic. You may want to copy and paste in the player or other objects to show some of the game, as well as some text to name/describe it. You can add some text, "Click to Start", to prompt the player.

Before we code our new *title* sprite, we need to change some things in some of the other sprites. In *MovePlayer*, we'll need to both remove and add some code. Start by removing the ❶ •**[Set ["*Level*"] to (*1*)]**, •**[Set ["*Play*"] to (*True*)]** and •**[Broadcast ["*NewLevel*"]]**. These will be handled by the title after this step. Then to the //*New Level* stack at the bottom, add a ❷ •**[Show]**, and in the //*Initialize* stack, start it with a ❸ •**[Hide]**. Now, since the player is actually drawn by the *draw player* sprite, we'll need to tweak its code as well. Add a ① •**[Hide]** at the top of its *//Hurt Countdown* to start it hidden. Then add a ② •*"NewLevel"* event, and under it a •**[Wait (1) Seconds]** and •**[Show]**. This should get your player hidden until the game starts.

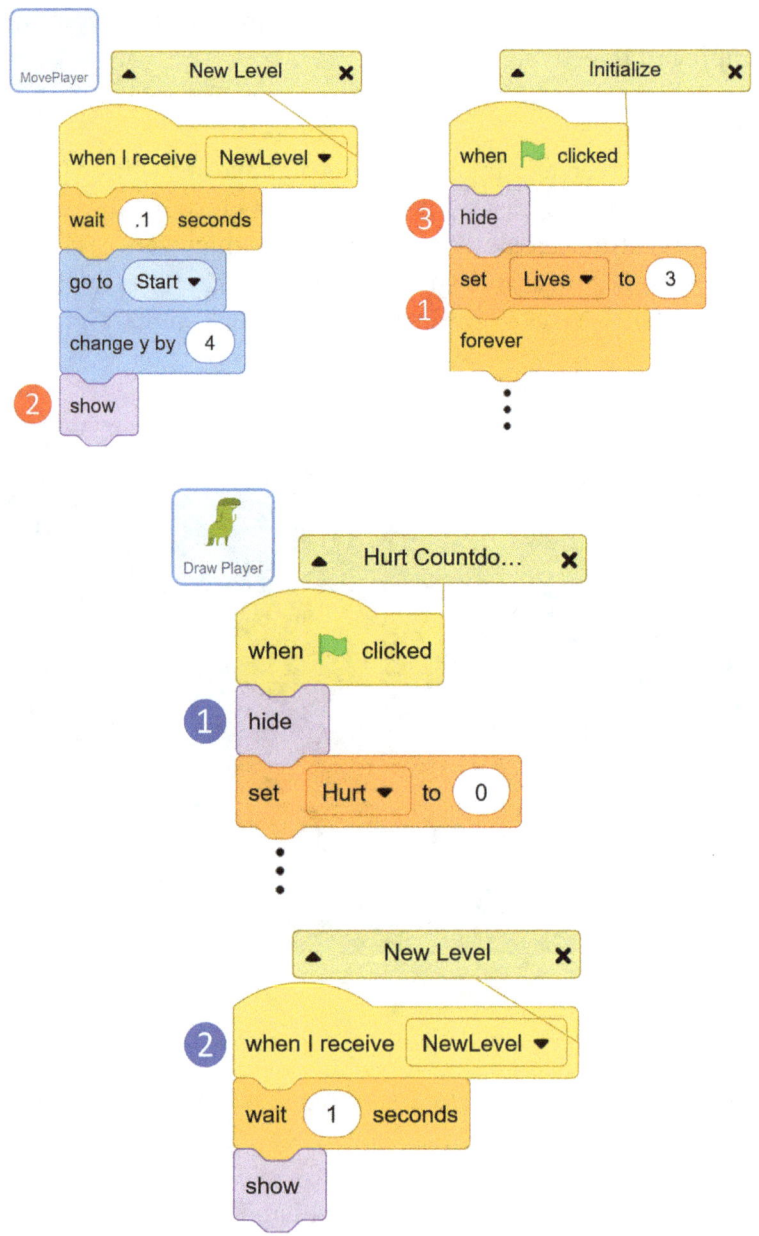

In the code for the *title*, have a ① •**[When ▷ Clicked]** that will •**[Set ["Play"] to (False)]**, •**[Go To [Front] Layer]**, and •**[Show]**. With the changes in the other sprites, all this will ensure the other sprites aren't active and that this screen covers everything else.

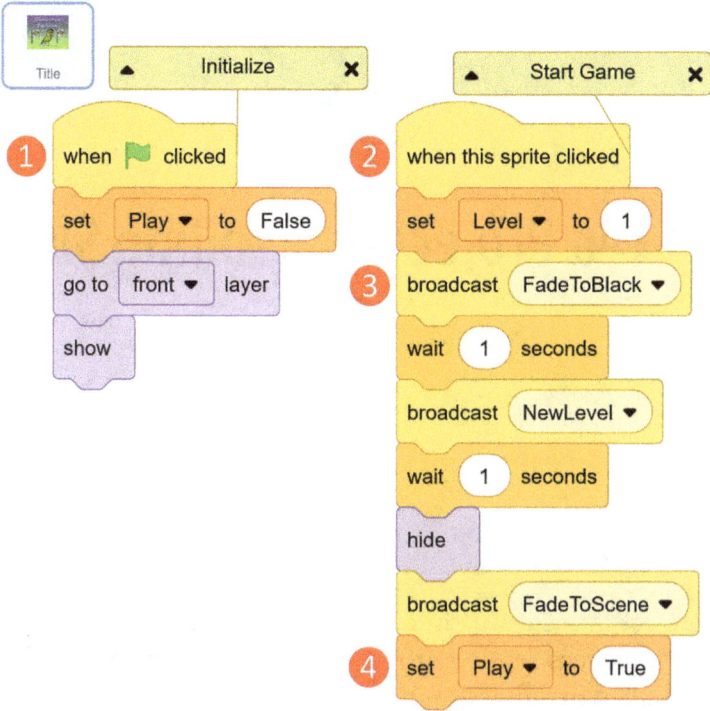

In a ❷ •[When This Sprite Clicked], put a •[Set ["*Level*"] to (*1*)] to set the starting conditions for the game. To create some cinematics, follow this with a ❸ •[Broadcast ["*Fade to Black*"]], •[Wait (*1*) Seconds], •[Broadcast ["*NewLevel*"]], •[Wait (*1*) Seconds], •[Hide], and then •[Broadcast ["*Fade-toScene*"]] and then ❹ •[Set ["*Play*"] to (*True*)] to start play. This will fade out the title, let the level load behind the *fader*, and then fade in to reveal the game. A nice, dramatic opening.

We've now got a working platformer game with examples of a lot of great mechanics for you and your class to explore and implement in your own custom creations! You learned a lot of useful professional tricks and tactics making this. Only one more project to go!

For working copies of this and every project in the book series, visit www.massivelearning.net for direct links to Scratch Projects, and to see our other projects and resources for coding education!

8

Advanced Project 4: Scrolling Shooter

What This Project Is

Our final project launches us into space for an arcade classic, the Scrolling Shooter. Here, players will control a spaceship flying through space while blasting away alien invaders. We'll have lots of action with waves of different types of enemy clones. Health levels and graphics not only show destruction animation but damage states as well. Work the way through battles to face off against a boss, a larger, more powerful enemy, to win the game. Additionally, multiple difficulty levels will be available for a range of play experiences. In around three hours (for adults), we should be able to build our basic Scrolling Shooter, with plenty of options for extending it.

What We're Learning with It

In our final project, we're taking a number of the basic principles and techniques already learned and extending them to show off a lot of professional techniques. Extensive cloning will generate wave after wave of enemies, even having clones creating clones! To bring our opponents to life, we'll create different AI movement patterns and boss modes. Revisiting stepcounts, currencies, and timers by adding in invulnerability states, ammunition, health states, rates of fire, and more. This project really helps take everything we've learned and reinforces it with practical game development example techniques.

Advanced Project 4: Scrolling Shooter ◆ 109

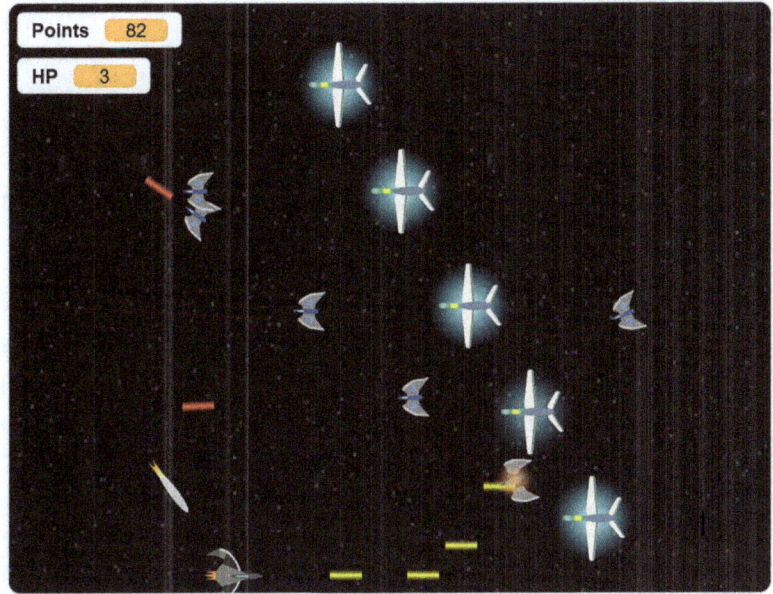

Movement Patterns. We'll learn to create movement sequences for enemy AI and switch between these states.
Scrolling Backgrounds. To create the illusion of movement, we'll create backgrounds that can scroll seamlessly, including a sneaky trick to move things off-screen!
Clone Properties vs. Parent Properties. Extensively managing the properties of both parents and individual clones and some handy techniques for dealing with clones separate from parents.
Clone Waves. How to generate waves of enemies, used to create different formations of bad guys.
Difficulty Selection and Scaling. A difficulty selector to alter the challenge of the game in quick and easy ways.
Procedural Generation. Generating enemies through code to create random challenges, making the game different every time you play.
Guide Objects. An invisible object that helps control other objects, a handy technique in Scratch.
Dynamic Spawning. Get our clones to create clones, which requires some interesting additional techniques to ensure proper relative positions and directions.
Cloud Variables. By accessing Scratch's cloud variables, we'll keep and display high scores from any player around the world, creating a sense of worldwide competition!

Building It

Step 0: Create Your New Project
Make sure you're logged in to Scratch, then click Create to begin a new project! Since we won't be using it, we can delete the Scratch Cat sprite by clicking on the trash bin on that sprite's thumbnail in the Sprite Listing. We can set the **backdrop** to just a plain black rectangle for now.

Step 1: Player Basics
With the game set in space, we won't have to worry too much about terrain like in our last project. Start by making the player. Create a blank sprite, *"Player"*, and make a spaceship design (I actually repurposed the **Dinosaur3** sprite in the example). If you do more than one costume, you can make a flame exhaust or running lights to animate like in the example. This helps bring the spaceship to life a little more and helps add to the illusion of movement. We'll create a more involved background later, but testing things on a solid black background will be good visual feedback for making sure our sprites look good and stand out against the darkness.

For code, keep in mind we'll build a menu system. ① Start our *player* with a •**[Hide]** and a •**[Broadcast ["GameStart"]]**. The game start event will be moved to the menu when it is done. In a ② •**"GameStart"** event, we can set some parameters. Declare a hidden variable •*"Speed"* by setting it to 6. This will be the rate the player can move in the game. We can choose a starting position, generally the middle left side of the screen, with a •**[Go To X: (#) Y: (#)]**, scale the player to something appropriate with a •**[Set Size]** (you'll need this for later even if you just use 100%), and •**[Show]**. This stack will just be used to set variables. Another variable will be used to handle movement, helping split our code into more manageable and readable stacks.

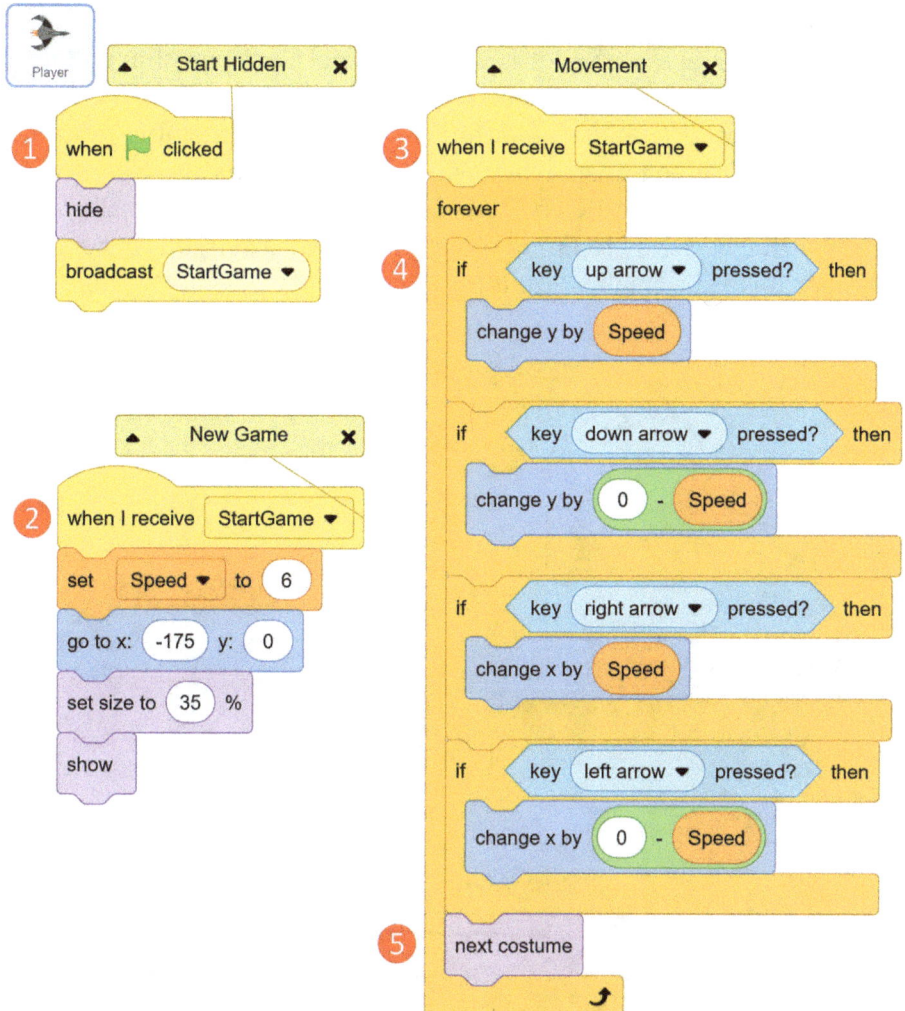

In another ③ •*"GameStart"* event, add a •[**Forever**] and key press tests. In this case, I only used the arrow keys, since we'll be using some letter keys for other commands, but you could add in an alternate WASD control scheme like in the Platformer project. ④ In each directional button, use •*"Speed"* or •((0)-•(*"Speed"*)) to move the player the appropriate amount of X/Y. If you have multiple costumes to animate, end the •[**Forever**] loop with a ⑤ •[**Next Costume**] to get a very rapid flicker to the exhaust flames. If you use a different style of animation, you might need to use a stepcount system like we did in the earlier Point-and-Click Adventure game to slow down the animation without affecting the controls (or split it to a separate animation •[**Forever**] loop of its own).

Step 2: Fire Lasers!

With our *player* now moving, let's get them firing! Our spaceship will fire out lasers on a button push, so we need to create a blank sprite, *laser*, to shoot. A simple yellow rectangle will suffice. If you want something fancier, use gradients of yellow to transparency with two adjacent rectangles, one on top of the other, with one vertically flipped to make a glowy laser. Play-testing the game is the best way to figure out what a good size is, but generally you want to err toward a smaller size for challenge.

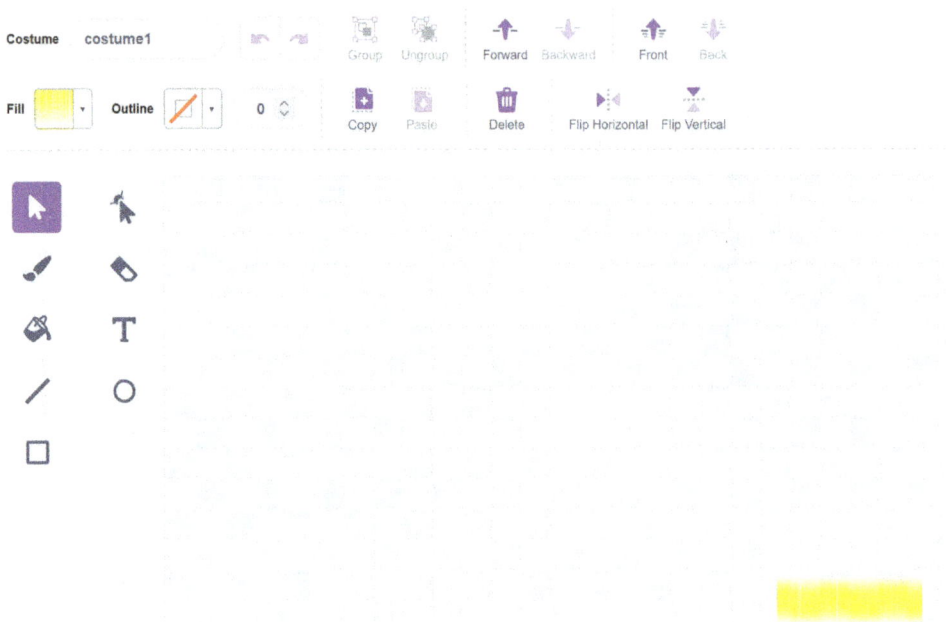

In the *laser's* code, ① start with a •[Hide]. ② •[When I Start As A Clone] will handle the general functionality. The *laser* needs to start with •[Go To [*Player*]] to make sure it's being generated in the right place. Then, •[Point In Direction (*90*)] and •[Show] so it will shoot right. Next, add a ③ •[Forever] loop, and •[Move (*10*) Steps] will determine its speed. Test if it shoots off-screen with a ④ •[If •<•(X Position) > (*239*)> Then], where we'd •[Delete This Clone]. This makes sure we don't end up with an ever-growing list of lasers flying infinitely through space off-screen. Creating more than 300 clones tends to crash Scratch, so you always want to make sure you're cleaning up after yourself and deleting clones not being used anymore.

Advanced Project 4: Scrolling Shooter ◆ 113

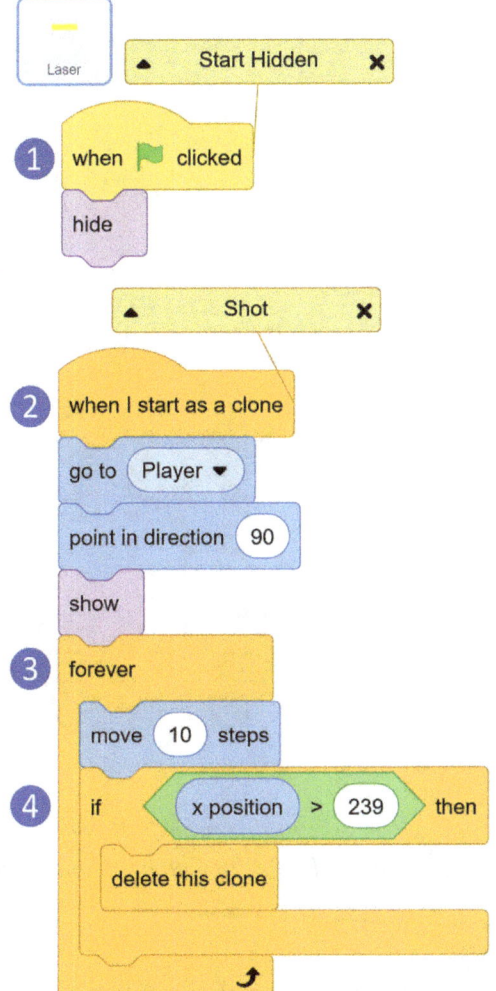

To fire our *laser*, go to the *player* code. Add a new variable to our •*"Game-Start"* variable stack with a ① •[**Set ["***ROF***"] to (*0*)**]. •*"ROF"* is a common acronym in game development that stands for "rate of fire", or how quickly you can use something again (also called a "cool down"). Create a new ② •*"GameStart"* stack for handling weapons. In a •[**Forever**] loop, add a ③ •[**If** •<•("*ROF*") < (*1*)> **Then** {} **Else** {}]; this will only run when the •*"ROF"* timer has cleared. Inside it, check if the Z key is pressed (the player is choosing to fire the laser), and inside it place a •[**Create Clone of [***Laser***]**] and a •[**Set ["***ROF***"] to (***5***)**]. In the •**Else** {}, place a ④ •[**Change ["***ROF***"] by (-***1***)**]. We now have a timer system to make sure the player can only shoot every five frames at most.

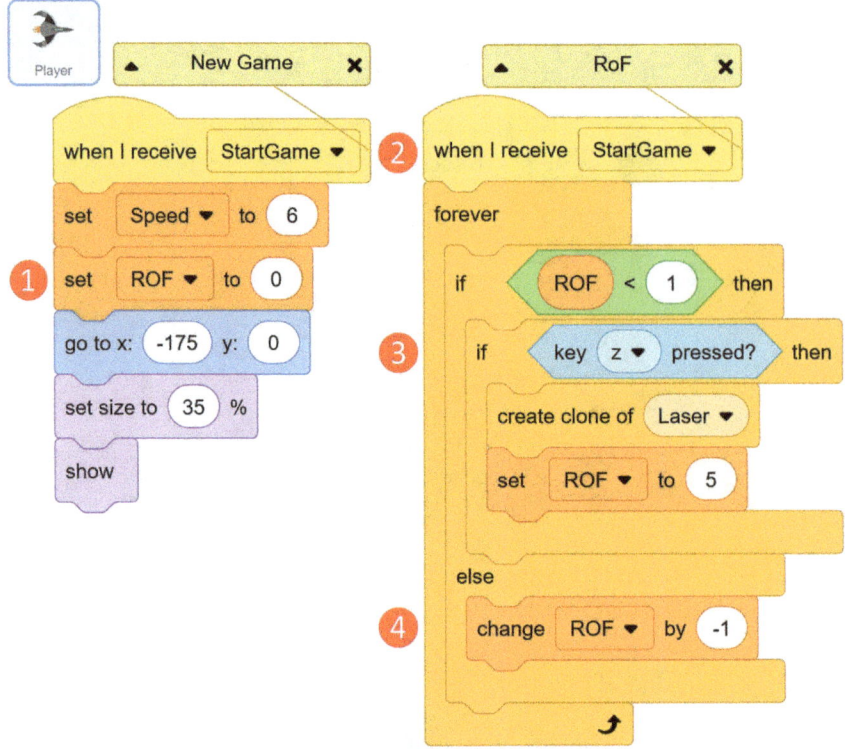

We'll be using clones extensively in this project. If you need a simpler primer for them, you can check out the Butterfly Catcher game in Book 1: Beginner!

Step 3: Bomber Enemies

Now that the *player* can shoot, we can start adding some opposition. Part of our enemy generation and AI system is a guiding object. This *pointer* sprite will be used as a reference position for generating and moving enemy units. Start by adding the **Arrow1** sprite, renaming it *Pointer*, and setting its **direction** to 0. In the code, add a ❶ •*"GameStart"* event with a •[Forever]. Inside it, use a •[Move (5) Steps] and then add an ❷ •[If <•(Y Position) < (-179) Then] with a •[Turn Right (180) Degrees] inside. Add a similar •[If <•(Y Position) > (179)> Then] with another •[Turn Right (180) Degrees] inside. Test it out and you should see the *pointer* bouncing up and down from the top to the bottom of the screen. Now that we know it's working, we'll add a ❸ •▶, [Hide], •[Go To X: (-240) Y: (0)] and •[Point in Direction]. These will set it off-screen behind the *player*. We want it to be past the edge of the screen to guide enemies to a final destination beyond the area of play. Because we're outside the screen, we can't use the [If On Edge, Bounce] code block, because it's always past the edge, so we needed our pair of •**Ifs** to do the job instead.

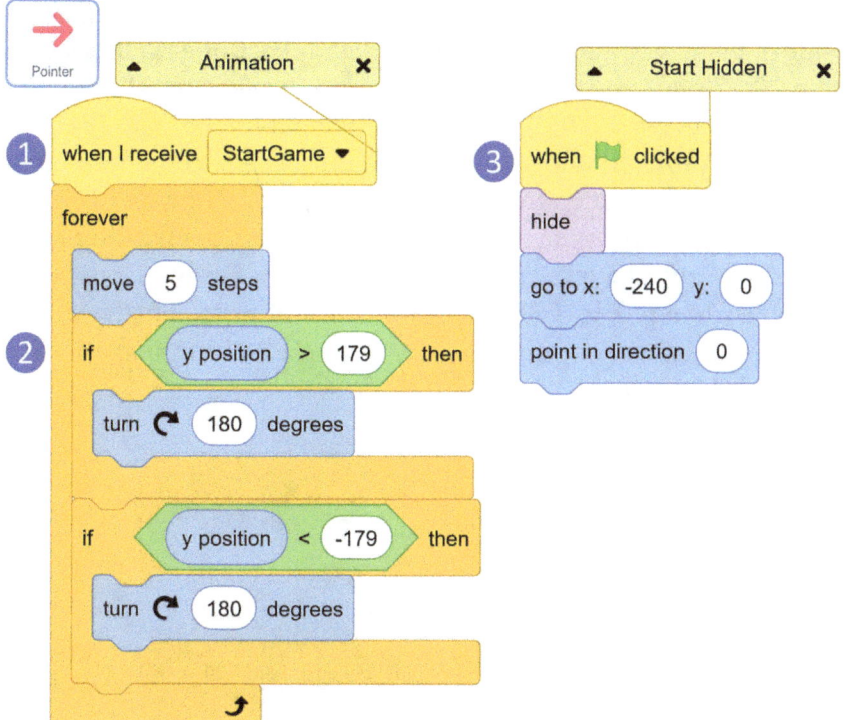

With the *pointer* bouncing away, we can now add our *bomber*. Add a blank sprite, *"Bomber"*, and draw in your own design for a *bomber* or repurpose and rename the **dragonfly** sprite like I did. Add four costumes for our *bomber* by duplicating its base design four times. In the first costume, add a strong force field. I suggest using a circle big enough to contain the whole sprite, with no outline and a fill of bright cyan to transparent radial gradient, then push it to the **back**. You can copy and paste this force field into the second sprite, but change the colour to a darker cyan and again push it to the **back**. The third costume needs no changes. In the fourth, we want a destroyed version of the *bomber*. You can break it apart, tilt parts, and add an explosion, and also copy and paste the force field and make it bright orange to be a fiery explosion cloud, and push it to the **back**. I renamed the **costumes** to tell them apart in the code easier. The first **costume** is *"3HP"* ("HP" stands for "hit points" or the amount of damage something can take before it is destroyed in game terms), second *"2HP"*, third *"1HP"*, and fourth *"Destroyed"*.

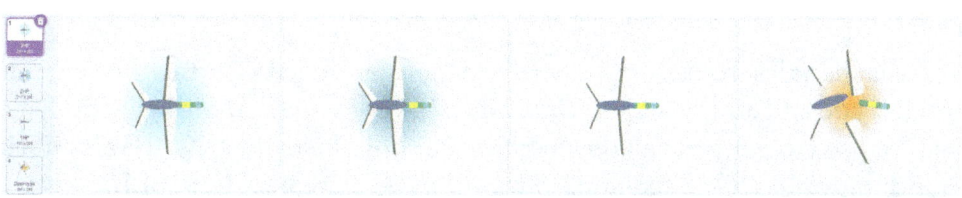

In our *bomber's* code, ❶ start with a ●[Hide] and an ●X position past the right side of the screen and make sure it begins as costume *"3HP"*. ❷ ●[When I Start As A Clone] and set the ●Direction to -90 so it goes left toward the player and ●[Set X to (200)] to start just visible on-screen. ❸ ●[Set Y to ●([Y Position] of [*Pointer*])] allows us to have our *bomber* clone to a constantly changing position, so when they get generated, they'll spread out and create banking groups to sweep the screen. Then, ●[Set Size to (#)] to an appropriate size, ●[Go To [*Front*] Layer], and ●[Show]. In a ❹ ●[Forever] loop, set a basic speed with ●[Change X by (-3)] to make it constantly scroll to the left, closing, then passing, the *player*. Lastly, use a ❺ ●[If ●<Touching (*Edge*)?> Then] to ●[Delete This Clone].

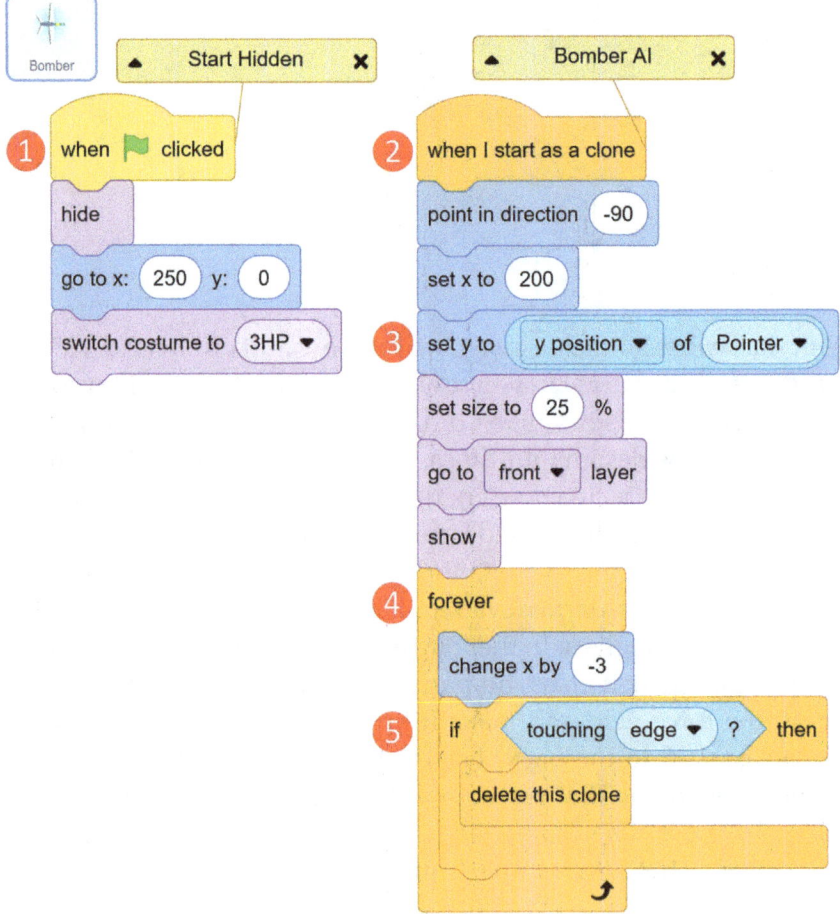

That handles the *bomber*, but how do they get cloned? For this, head to our *stage's* code and add in a ❶ ●*"GameStart"* event with a ●[Forever] loop. This will be used to generate all the enemies. For now, just add a ❷ ●[Wait ●(Pick Random (6) to (10)) Seconds], followed by a ●[Create Clone of [*Bomber*]].

Now we have an endless string of bombers flying by you can play-test with. Also, add a ③ •*"Lost"* event and •**[Stop [***Other Scripts In Sprite***]]**; soon we'll add a Game Over that will need the enemies to stop being generated.

Step 4: Damage and Destruction

You probably have noticed pretty quickly that we can't shoot the *bomber*, nor does the *bomber* collide with the *player*. That's because we haven't added any code to damage or destroy anything yet. Let's start with the simple task of adding collisions to the *bomber*.

First, add a new ① •**[When I Start As A Clone]** event with a •**[Forever]** loop to handle damage. In it, test ② •**[If** •<•**<Touching (***Player***)?> or** •**<Touching (***Laser***)?>> Then]**, so it will suffer damage when shot or on colliding with the *player*. Inside it, add a SFX **Crunch**. We'll have to add that in the **Sound** tab first, then add a ③ **[Next Costume]** to cycle through the damage visuals for the sprite.

Next, add another ④ •**[When I Start As A Clone]** event with a •**[Forever]** loop for handling its final destruction. Here, we test ⑤ •**[If** •<•**(***Costume Number***) = (4)> Then]** to check if the *bomber* has switched to its destroyed graphic. Add a new variable •*"Points"* to track the player's score. Inside the conditional, add a •**[Wait (***0.1***) Seconds]**, ⑥ •**[Change [***"Points"***] by (5)]**, and then •**[Delete This Clone]**. This will allow the clone to be destroyed and give the player *"points"* but also adds a delay so the destruction is visible, rather than instantaneous.

118 ◆ Advanced Project 4: Scrolling Shooter

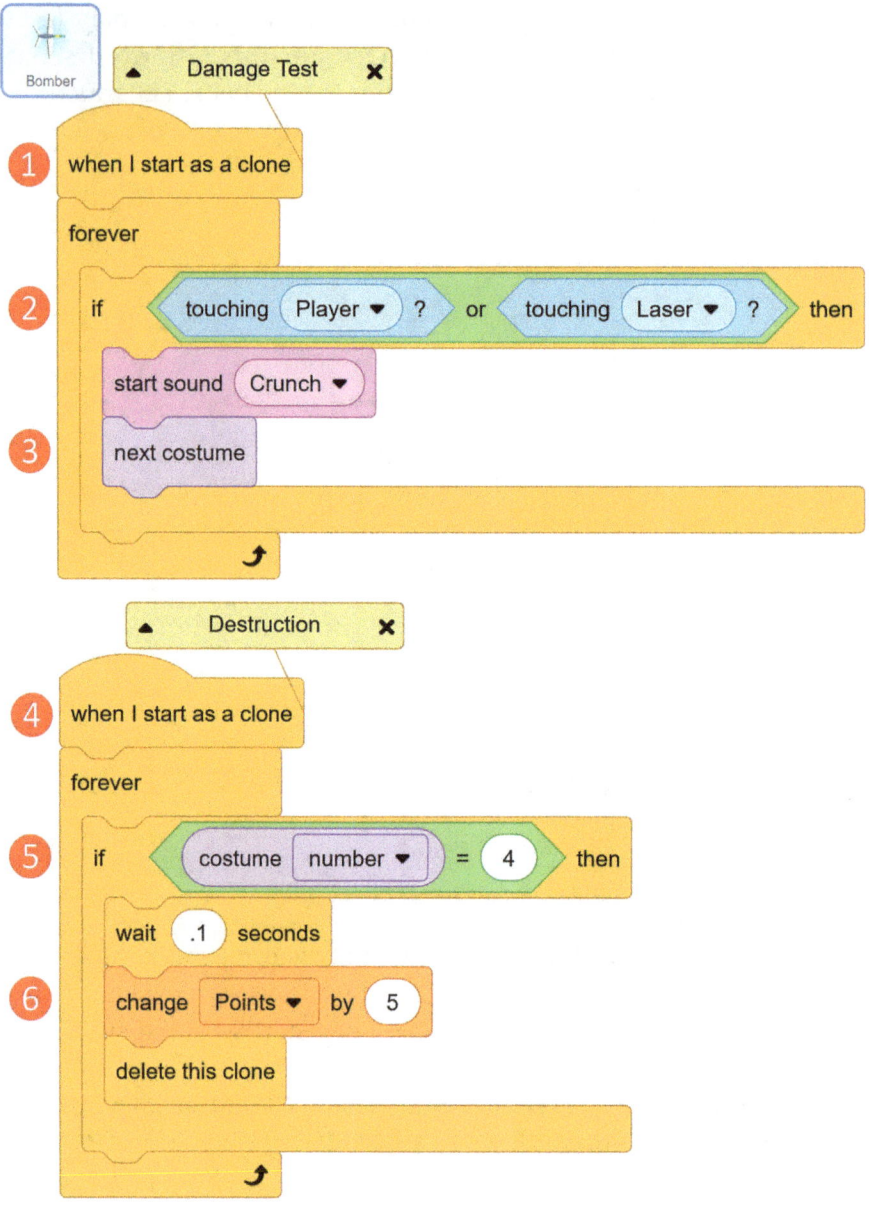

Lastly, we'll update the *player*. In the •*"GameStart"* variables stack, ① add a •**[Set [*"Points"*] to (0)]** and add a new visible variable •*"HP"* to track the player's health. Add a •**[Set [*"HP"*] to (5)]** to the same stack. Then, add a new ② •*"GameStart"* event to track damage. We'll need a •**[Forever]** loop and test ③ •**[If •<Touching (*Bomber*)?> Then]** and add a •**[Change *"HP"* by (-1)]**. Now our *player* loses •*"HP"* when they collide with an enemy. It's far from perfect, but it's a start.

Advanced Project 4: Scrolling Shooter ♦ 119

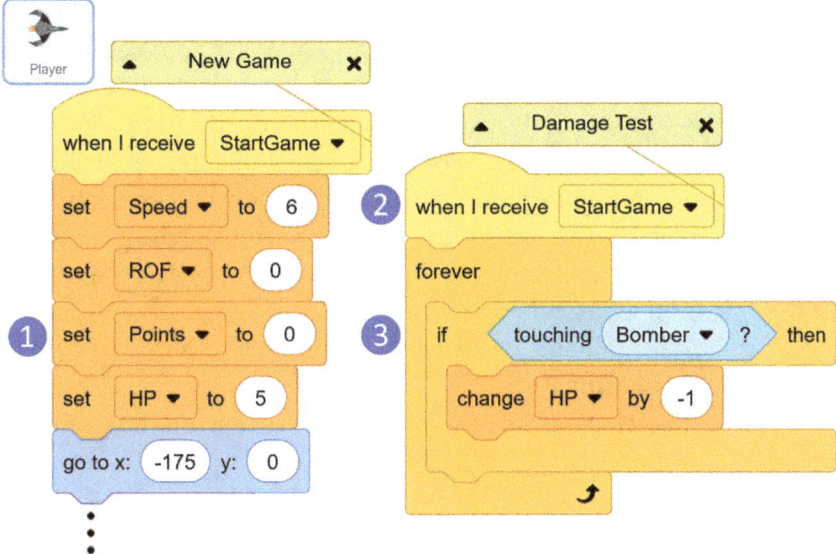

Step 5: Player Health and Death

While we've got the players HP counting down, it's not really an adequate health system. Let's improve on it, making the damage more visible and having it actually impact play. For this, create a new blank sprite "*HitFX*", which will be a graphical effect that displays whenever the *player* takes damage.

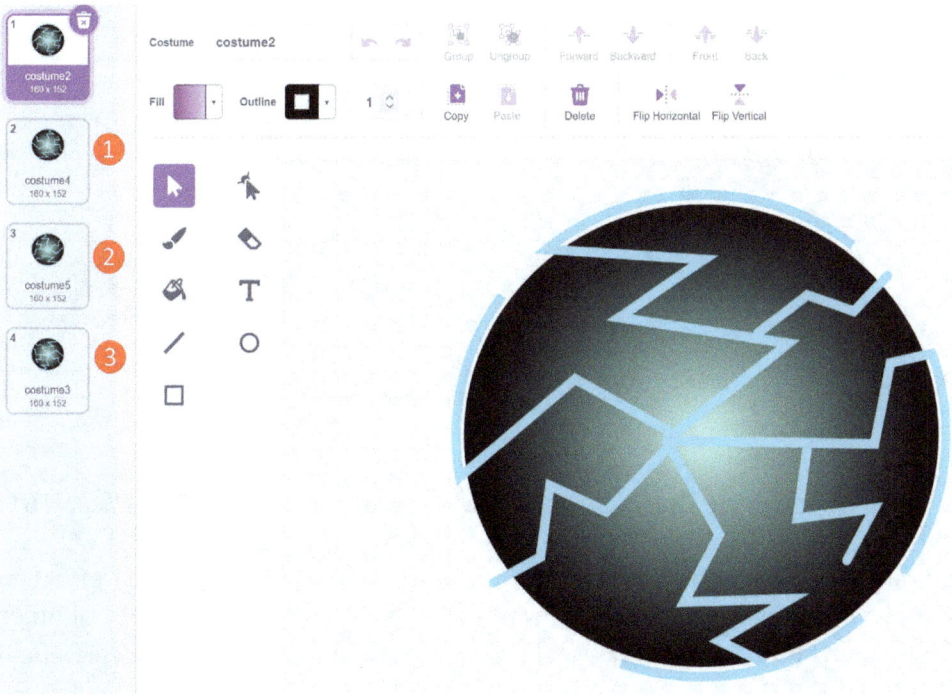

I suggest making a circle large enough to fit the ship with a bright cyan colour fading to black with a radial gradient. Then, set your outline colour to bright cyan and thickness to 10. First, make a new empty filled circle sized to match the edges of the circle, move it aside so you've got room, then erase most of the circle, leaving some random patches of lines around the edge. Then, make some random lightning bolt crackle lines dividing the circle with connected jagged patterns, attaching the remaining circle portions. Move these lines back over the main circle. Now you've got a cool shield crackling with energy design! Copy this costume three times, then ① in #2 select it all and then flip horizontally. In costume #3, ② select all and flip vertical and flip horizontal. In costume #4, ③ select all and flip vertically. This will create a cool crackling electricity animation for a force field struggling with a hit.

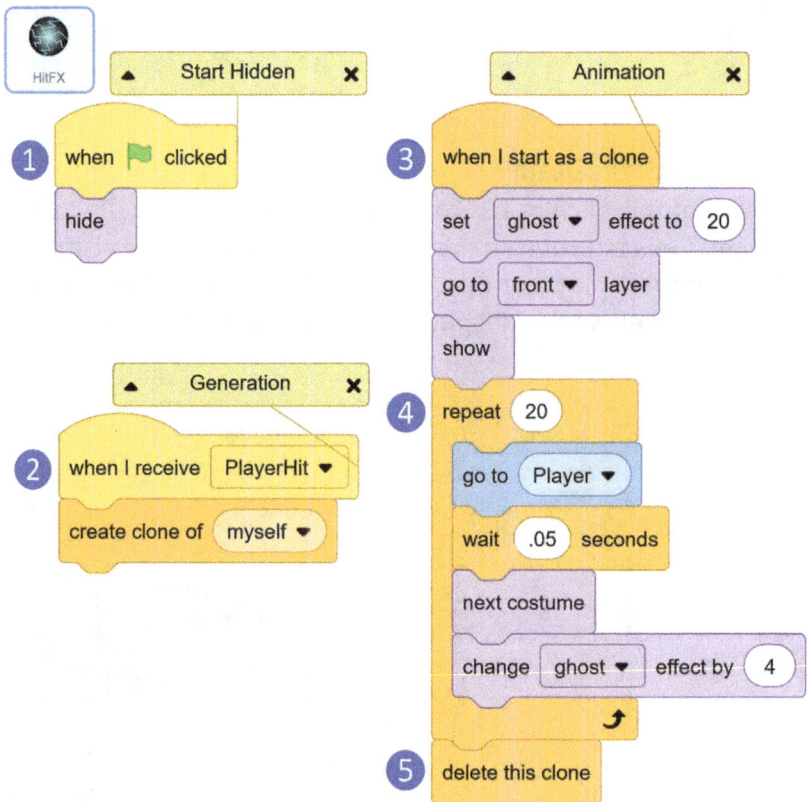

For its code, we'll ① start with it being hidden. In a ② •*"PlayerHit"* event, •[**Create Clone of [***Myself***]**] so whenever the player is hit, it will create a copy. ③ •[**When I Start As a Clone**] should •[**Set [***Ghost***] effect to (20)**] so the shield will be transparent to view it over the player but not block them, then •[**Go To [***Front***] Layer**] and •[**Show**]. Next, we need a ④ •[**Repeat

Advanced Project 4: Scrolling Shooter ♦ 121

(20)] to animate it, and •**[Go To [Player]]** so it can always stay with the *player* even as they move, •**[Wait (.05) Seconds]**, •**[Next Costume]**, and •**[Change [Ghost] effect by (4)]**. This will have the shield crackle but fade away. After the •**[Repeat]**, ❺ •**[Delete This Clone]**. Set HitFX's size so it just covers the player.

Now, switch to the *player's* code. We'll need a new hidden variable, *"Invulnerable"*, and to add a ❶ •**[Set ["Invulnerable"] To (0)]** in our variable, initializing in our //*New Game* stack. In our //*Damage Test* stack, add a ❷ •**[If •<•("Invulnerable") = (0)> Then]** test and move our touching bomber conditional inside it.

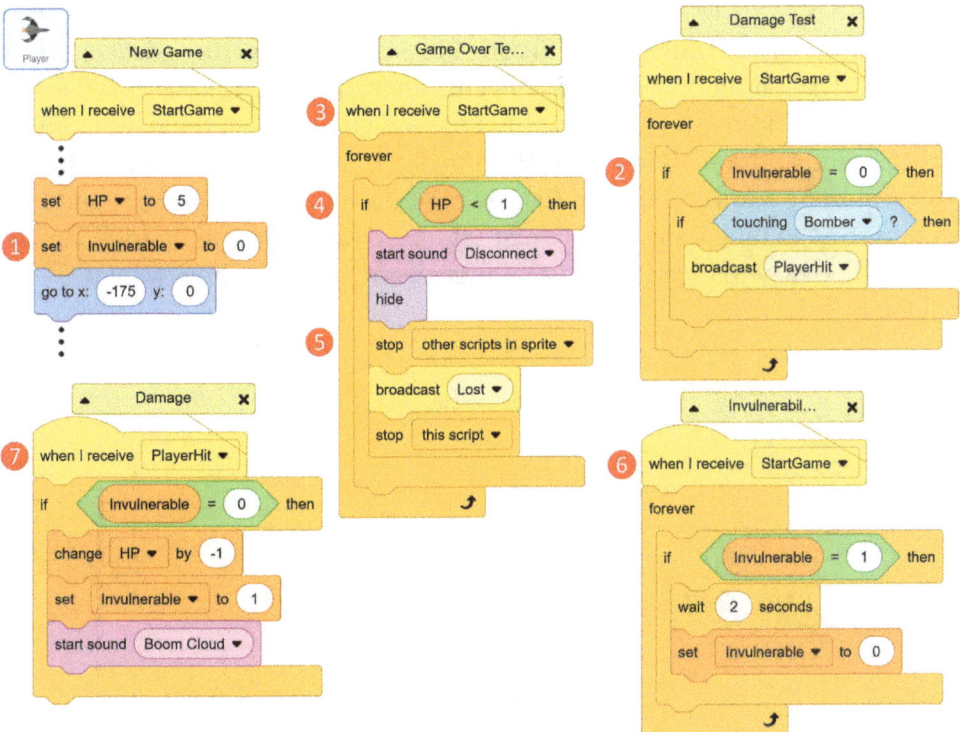

Next, add a few more stacks. In a ❸ •*"GameStart"*/•**[Forever]** loop, we'll test ❹ •**[If •<•("HP") < (1)> Then]** to handle a Game Over. Inside it, add a •**[Start Sound [Disconnect]]** (after adding the sound in the **Sounds** tab), •**[Hide]**, ❺ •**[Stop [Other Scripts In This Sprite]]**, •**[Broadcast ["Lost"]]**, and then •**[Stop [This Script]]**. By stopping other scripts, the player will no longer be responsive, and •*"Lost"* will allow us to stop other objects, and then •**[Stop [This Script]]** shuts down this test so it doesn't run again. Since all the code is within •*"GameStart"* events, though, they can be restarted with a new game, but we'll get to that with our *menu* later on.

In another new ⑥ •"*GameStart*"/•[**Forever**], we'll handle our invulnerability state. •[**If** •<•("*Invulnerable*") = (1)> **Then**] tests if the player is invulnerable, •[**Wait (2) Seconds**], and •[**Set** ["*Invulnerable*"] **to (0)**]. This will detect when the *player* becomes invulnerable and, two seconds later, make them vulnerable again. This feature, a short-term invulnerability, is a common feature in games that prevents the *player* from being hit by something each frame, therefore limiting them to losing only 1 HP to each mistake they make (each time they collide with something). Without a system like this, colliding with things that cause damage but don't immediately get destroyed would cause damage 30 times a second! So this gives the *player* some grace and improves the playing experience.

Add a ⑦ •"*PlayerHit*" event. Inside the •[**If** •<**Touching (***Bomber***)?**> **Then**], delete the •[**Change** "*HP*"] code block and add in •[**Broadcast** ["*PlayerHit*"]]. Then, in the •"*PlayerHit*" event, add •[**If** •<•("*Invulnerable*") = (0)> **Then**], which limits the *player* to only taking damage when they aren't invulnerable, and add our •[**Change** ["*HP*"] **by (-1)**], •[**Set** ["*Invulnerable*"] **to (1)**], and •[**Start Sound** [*Boom Cloud*]] (you may want to soften, shorten, and fade out this SFX to tone it down). We just set up a distinct way to handle damage that will be easy to extend to other dangers, and a Game Over system.

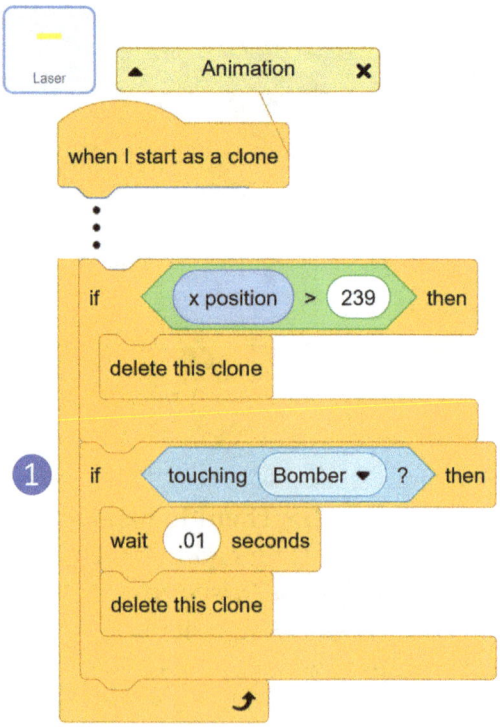

One last thing to change is the *laser*. You'll notice that when you shoot the bomber, the laser just keeps going straight through it, and because it collides step after step, just a single laser shot will destroy the *bomber*. So in our *laser* code, let's make something that'll correct that. In our •[**Forever**] loop, add a conditional ① •[**If** •<**Touching** (*Bomber*)?**> Then**]. Inside we'll •[**Wait (.01) seconds**] and then •[**Delete This Clone**].

> **Race Condition**
>
> *Why do we need a •[Wait] block here? This is here to ensure the laser doesn't instantly delete itself. If you remove it, you'll see that sometimes the laser detects the collision first, and if it deletes itself, it will be gone and the bomber won't trigger its collision with it. With the •[Wait] in, we don't have that problem, though the reverse can happen and the bomber could sometimes take two hits by detecting the collision first, then the laser delays destruction, triggering a second collision before it's destroyed. We'll just consider those "critical hits" and call it a day (a feature, not a bug)! In the computer, each sprite (and the stage) will run through their code in order, each frame of animation, or 1/30th of a second. Unfortunately, we don't know what order the sprites will run in. This means that one sprite might run its code and detect a collision but could then move or destroy itself, and then when the next sprite runs its code, the first sprite has moved or destroyed itself so the second sprite doesn't detect a collision and doesn't react. In more advanced code editors, we could access that information or control for it, but Scratch isn't meant for that level of deep digging, so here we use a wait command to ensure both react, but it does lead to the occasional double hit in the game. Alternatively, you could use an invulnerability state to prevent that.*

Step 6: Enemy Waves

For our next step, we'll improve how our enemies are generated. Start with the *stage* code. Delete •[**Create Clone of [***Bomber***]**] code block. In its place, use a new hidden variable •"*WaveType*" and under •[**Wait**], ① •[**Set ["***Wave-Type***"] to** •(**Pick Random (1) To (7)**)]. Then, under it add an ② •[**If** •<•("*WaveType*") = (#)> **Then**] for each result 1 to 7. Inside the 1, 4, 5, and 7 *Ifs*, add a •[**Broadcast ["***BomberWave***"]**] code block.

Now, switch to our *bomber's* code. Here, add a ① •"*BomberWave*" event. Inside it, put a ② •[**Repeat** •(**Pick Random (3) to (8)**)]. This will repeat the inside code, but do it for a random number from 3 to 8. Inside it, add a •[**Wait (0.5) Seconds**] and a ③ •[**Create Clone of [***Myself***]**]. Now, any wave that generates *bombers* will clone three to eight of them. For now, waves 2, 3, and 6 won't generate enemies, but we'll take care of that later.

124 ◆ Advanced Project 4: Scrolling Shooter

Also, let's include code to handle our Game Over system. Add a ④ •*"Lost"* event and add a •**[Delete This Clone]**. Also, do this ① for our *laser* sprite.

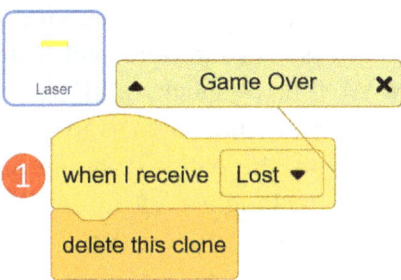

Step 7: Scrolling Background

To make our game look more like the player is flying through space, we'll be adding in a space background and make it move. Find a simple star field image, ideally just pure black with very simple stars dotted all over, so it can seamlessly repeat or tile without issue. Simple is good; you'll need something that you won't notice the edges changing, so avoid nebula along either left/right edge.

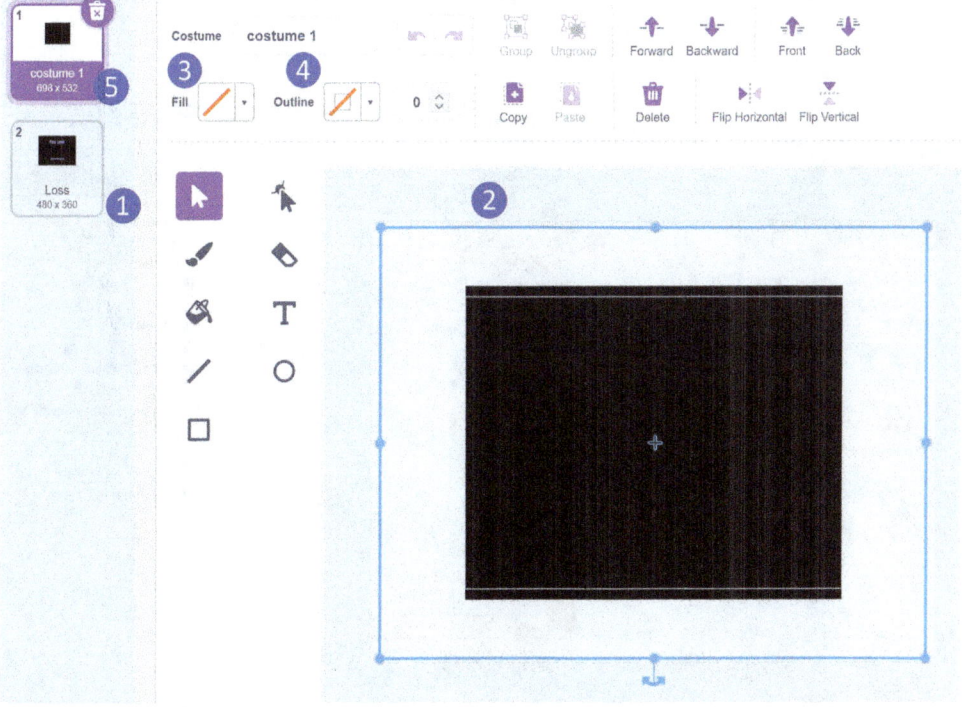

Use the **Choose A Sprite – Upload** option to add your image to your project. Scale it to fit the full Stage Window size. On the left-hand side, where it lists the costumes, make sure the costume dimension is exactly 480 pixels wide (① check the first number). Remember, Scratch has an off-screen protection system, so it will prevent objects from having too little of their image on-screen. Unfortunately, this will interfere with our background scrolling off the screen smoothly. We'll use a sneaky trick here to avoid that, the same technique we used for size restriction bypass in Bar Charts and Data Files. ② Make a rectangle that fills the entire drawing area, beyond the screen dimensions; now make the rectangle's ③ fill and ④ outline transparent. You might not see it, but that transparent rectangle and its ⑤ larger size will trick the off-screen protection, allowing our image to go off-screen enough for a smooth scrolling effect. Name this sprite "*SpaceBG*". We won't animate this sprite using costumes but instead use cloning and code to do it.

In a ① •"*GameStart*" event, •[**Switch Costume to** [*costume 1*]]. We'll be adding more costumes later, so might as well add this now. Set it to the back layer and position it at X: (0), Y: (0), then ② •[**Create a Clone of** [*Myself*]], because we'll need two copies of the background so we have one we're on and one we're moving to. Under that, add a •[**Forever**] loop with a •[**Change X by (-1)**] to create our scrolling motion. Then, we'll need an ③ •[**If** •<•(**X Position**) < (-479)> **Then**] to see if it moved completely off-screen. If so, it should •[**Go To** [*Back*] **Layer**] and •[**Go To X: (480) Y: (0)**] to position perfectly to the right of the other clone.

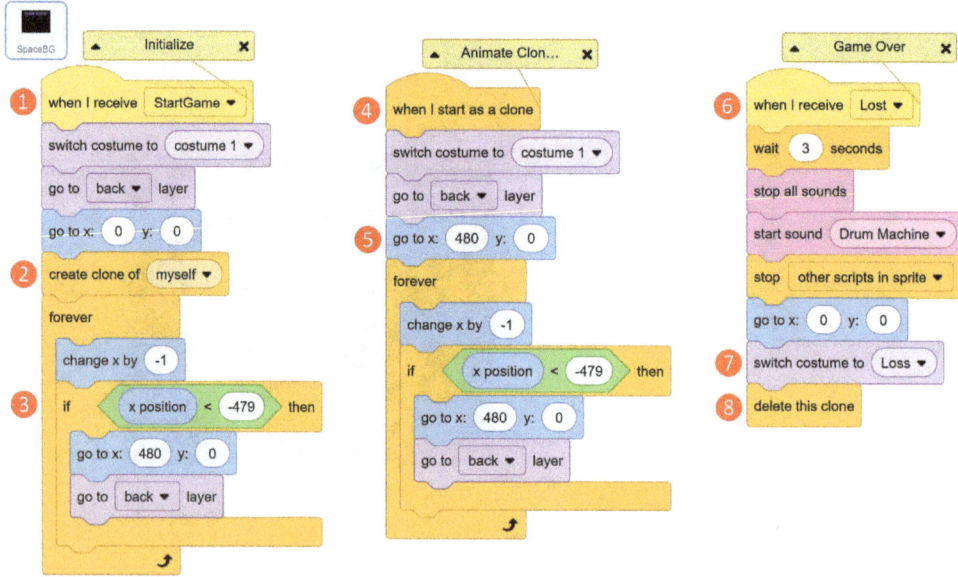

Advanced Project 4: Scrolling Shooter ♦ 127

Next, add a ④ •[When I Start As A Clone] and simply duplicate the code from the •"*GameStart*" event (and remove the create clone). The clone will be almost the same, but with one change. We'll want its original positioning to be ⑤ •[Go To X: *(480)* Y: *(0)*]. This begins it perfectly aligned to the right of the original. With them both moving 1 pixel/frame and resetting to X: 480, when they get to -480, they'll endlessly repeat.

Add in a ⑥ •"*Lost*" event. We'll •[Wait *(3)* Seconds], •[Stop [*All Sounds*]], •[Start Sound [*Drum Machine*]], •[Stop [*Other Scripts in Sprite*]], •[Go To X: *(0)* Y: *(0)*]], and add a new ⑦ costume and called "Loss". Duplicate the normal space BG but add text to indicate the loss or Game Over. We can even duplicate an enemy sprite on the centre of the image to rub the alien victory in the player's face. Finish up that stack of code by switching to that new "loss" costume. Finally, at the bottom of the stack, ⑧ you can •[Delete This Clone] so multiple game starts don't create additional clones.

We mentioned this technique as a way to spice up the Batty Flaps game in Book 1: Beginner, but you could also spice up this game with some kind of implementation of the gate obstacles from that game, so check it out!

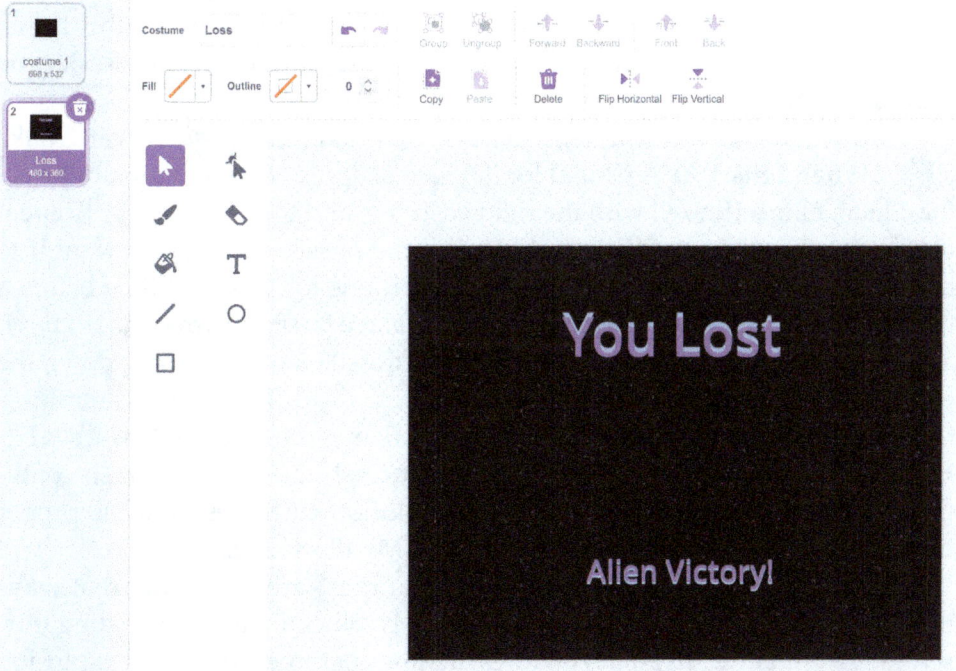

Step 8: Missiles

With our core game functioning pretty well, it's time to start adding more enemy types. Start by adding a missile, which will function as a stand-alone enemy but will also be a special attack launched by the *boss* later on. Add a blank sprite *"missile"*, and in the **Costumes** tab we'll be drawing a few costumes.

First, draw a simple missile. You can draw a long oval, then with reshape add an extra point on the left side and turn the two dots on the left side to **pointed** instead of **curve** types. Next, position them to make a rocket-shaped object. Add some flames coming out the back by reshaping a rectangle and setting its outline off and its fill to orange and yellow in a radial gradient. **Duplicate** that for a second frame and either lengthen or shorten the flame. That's a two-frame animation for the *missile!*

Next, we need to draw in two frames of explosions. Draw a circle and add a bunch more points and turn them to **pointed** types. Make it red, then copy and paste the shape, turn it orange, and rotate it to be offset from the first. This makes a nice cartoon blast kind of shape. You can duplicate this frame and simply shrink it and flip it to be the second frame of explosion. Resize the frames to make sense.

Now we can work on the code. Start by ❶ hiding the *missile* on start. Make a ❷ •[When I Start As A Clone] for its main action, and add in a sound effect for it launching, •[Show] with the right costume, and then have a ❸ •[Repeat (240)] loop. Don't use a •[Forever] loop for the missile because we want it to have limited range. Using a •[Repeat], it functions for a limited time before a hard-set limit, in this case 240/30 = 8 seconds. Have it •[Point Towards [*Player*]] and •[Move (4) Steps] so that it is a homing missile always chasing down the player, but moving slowly enough the *player* can evade it if they're careful. A • [Next Costume] code block will animate the flames, but we need an ❹ •[If • <•(*Costume Number*) > (2)> Then] •[Switch Costume To [*costume1*]] conditional to make sure it doesn't show the explosion costumes until the proper time. Below the •[Repeat], call a new custom My Block •["*Explode*"].

•"*Explode*" will detonate the missile, but there are many different ways it can happen, so we want to put this in a My Block to avoid repeating ourselves. ❺ •[Define "*Explode*"] with switching costumes to the first explosion

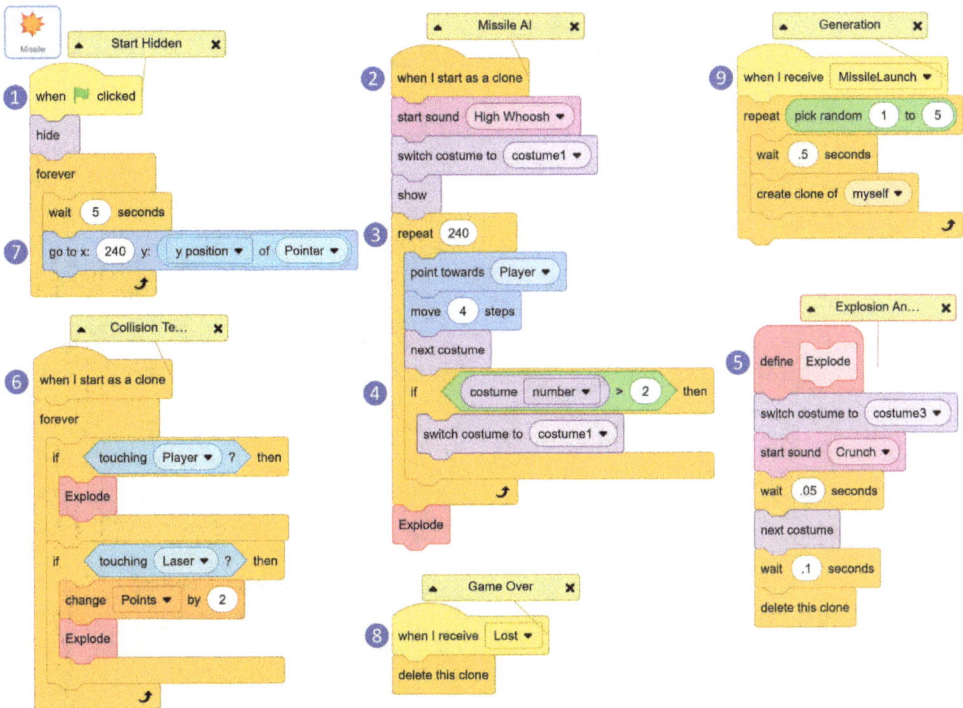

costume, playing a **Crunch** SFX, **[Wait (.05) Seconds]** then switching to the second explosion costume, **[Wait (0.1) Seconds]**, and then **[Delete the Clone]**. We use a simple two-frame animation to explode the *missile* (which, being larger than the normal *missile*, will potentially collide with the player even if the *missile* didn't hit directly) and then delete itself.

Add in another ❻ **[When I Start As A Clone]**/**[Forever]** combo to test for touching the player. If it does, it will **"Explode"**. We can duplicate this conditional and change it to test for touching the *laser*. This way, the *player* can shoot down the *missile*, ending the threat. If they do, they'll be awarded two **"Points"**. To make the missile generate from different positions, in the //Start Hidden stack, add a ❼ **[Forever]**, and inside place a **[Wait (5) seconds]** and a **[Go To X: (240) Y:** **([Y Position] of (Pointer))]**. Because that forever loop is in the ⚑ event, it only affects the original missile and won't reposition clones.

Next, we have two events to deal with. The first, ❽ **"Lost"**, will simply **[Delete This Clone]**. The second, ❾ **"MissileLaunch"**, will be used to spawn clones. Position the *missile* based on the Y position of the *pointer* along the right side of the screen. Add a **[Repeat (Pick Random (1) to (5))]**, and inside put a **[Wait (0.5) Seconds]** and **[Create Clone of [*Myself*]]**. Then, in the *stage* make sure our wave generation includes *missiles*. Add a ❶ **[Broadcast ["*Missile Launch*"]]** event to the 2, 4, 6, and 7 "*WaveType*" Ifs.

130 ◆ Advanced Project 4: Scrolling Shooter

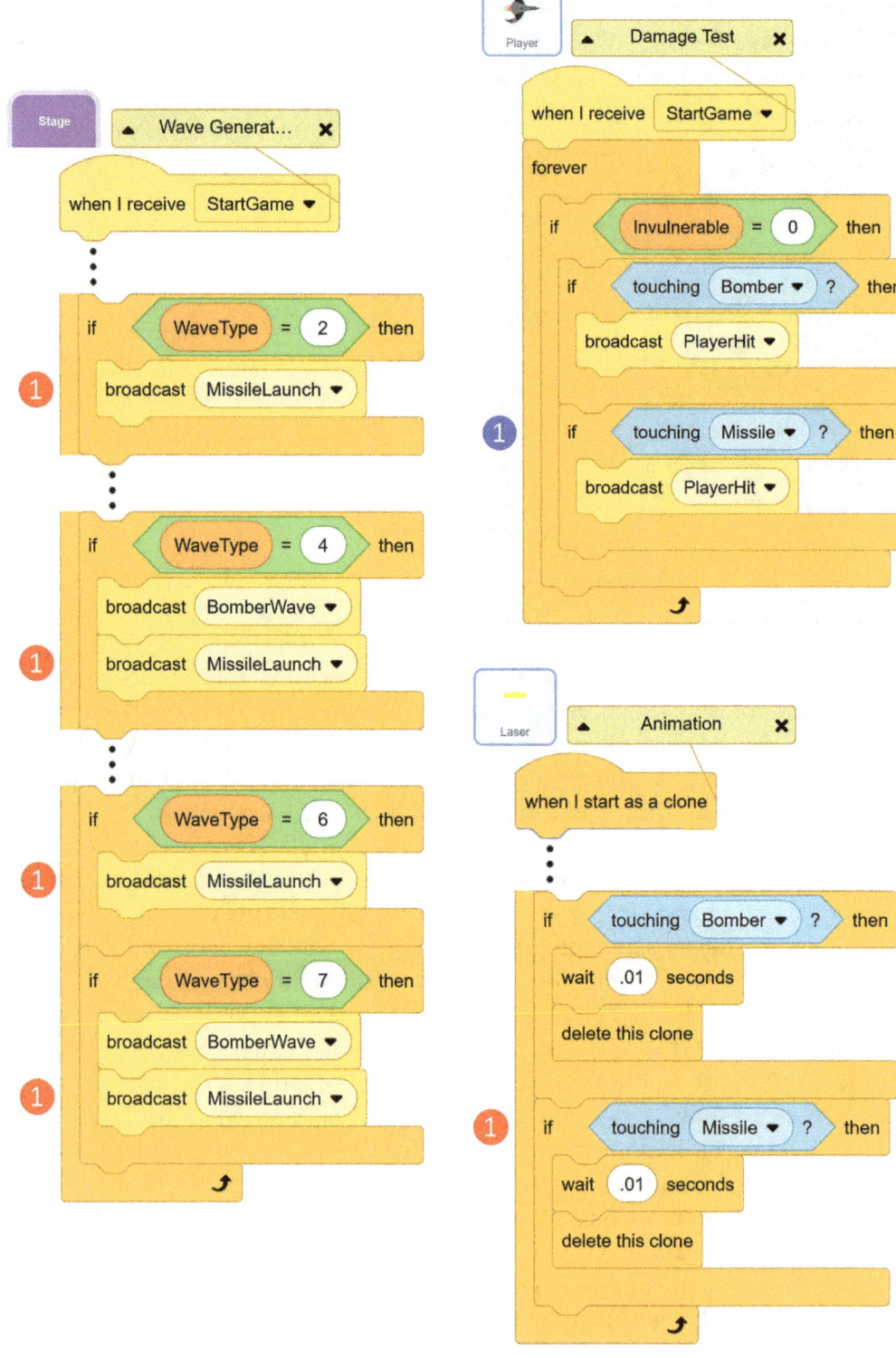

Lastly, update the *player's* code to get hit by *missiles*. In the *//Damage Test* stack, duplicate the *bomber* collision test *If* and change it to ❶ •<Touching [*Missile*]?>. Now, *missile* collisions will call •*"PlayerHit"* as well. Also, update our *laser* code and duplicate our •[If •<Touching [*Bomber*]?> Then] and set the duplicate to ❶ •[If •<Touching [*Missile*]?>Then].

Step 9: Fighter Enemies

Next, we'll add in a new type of spaceship enemy. This one will be more nimble and dangerous than the *bomber*, but also more frail. Start by duplicating our *bomber* as a base. For costumes, make a smaller and simpler *fighter*-style spaceship. We won't need multiple HP worth of costumes; just have a single flying state costume and a single destruction state costume. ❶ Make sure you change the starting costume to the new undamaged *fighter* costume.

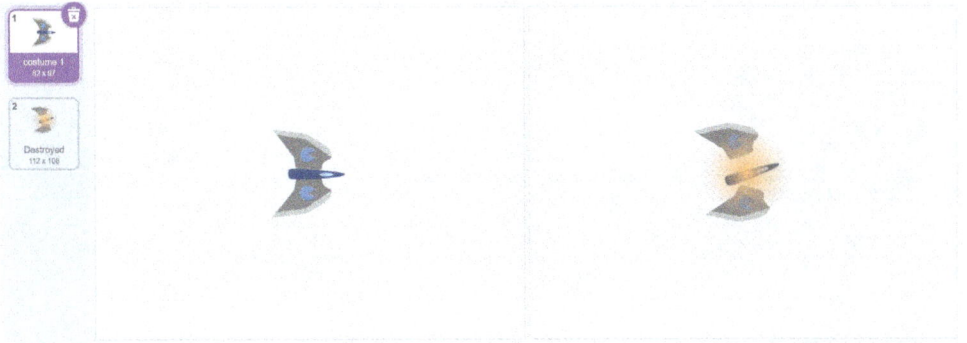

In the ❷ *//Fighter AI* stack code, switch •[Set Y to •([Y position] of [*Pointer*])] to •[Set Y to (Pick Random (-170) to (170))], then add a •[Point Towards [*Pointer*]] above the •[Forever] loop. The *fighter* will fly around at an angle rather than just straight across, but unlike the missile, it doesn't update its direction. You'll need to switch the •[Change X by (#)] to a ❸ •[Move (5) Steps] so they can move at an angle. By directing them to the *pointer*, their orientation will constantly change between clones and waves. We can also change the damage handling stack. Change it to use ❹ •<•(Costume Number) = (2)> and •[Change ["Points"] By (1)]. The generating event also needs to be changed. Rename it to ❺ •*"FighterLaunch"*, and change the pick random values to 3 to 8.

Since we've got a new enemy to collide with, the ❶ *player* and ❶ *laser* need to be updated like we did for the *missile*. Just duplicate the existing code and use it for *fighter*, or add in an •<<condition> or •<Touching (*Fighter*)>> to the appropriate stacks, where you see a •<Touching [*Bomber*]?>.

132 ◆ Advanced Project 4: Scrolling Shooter

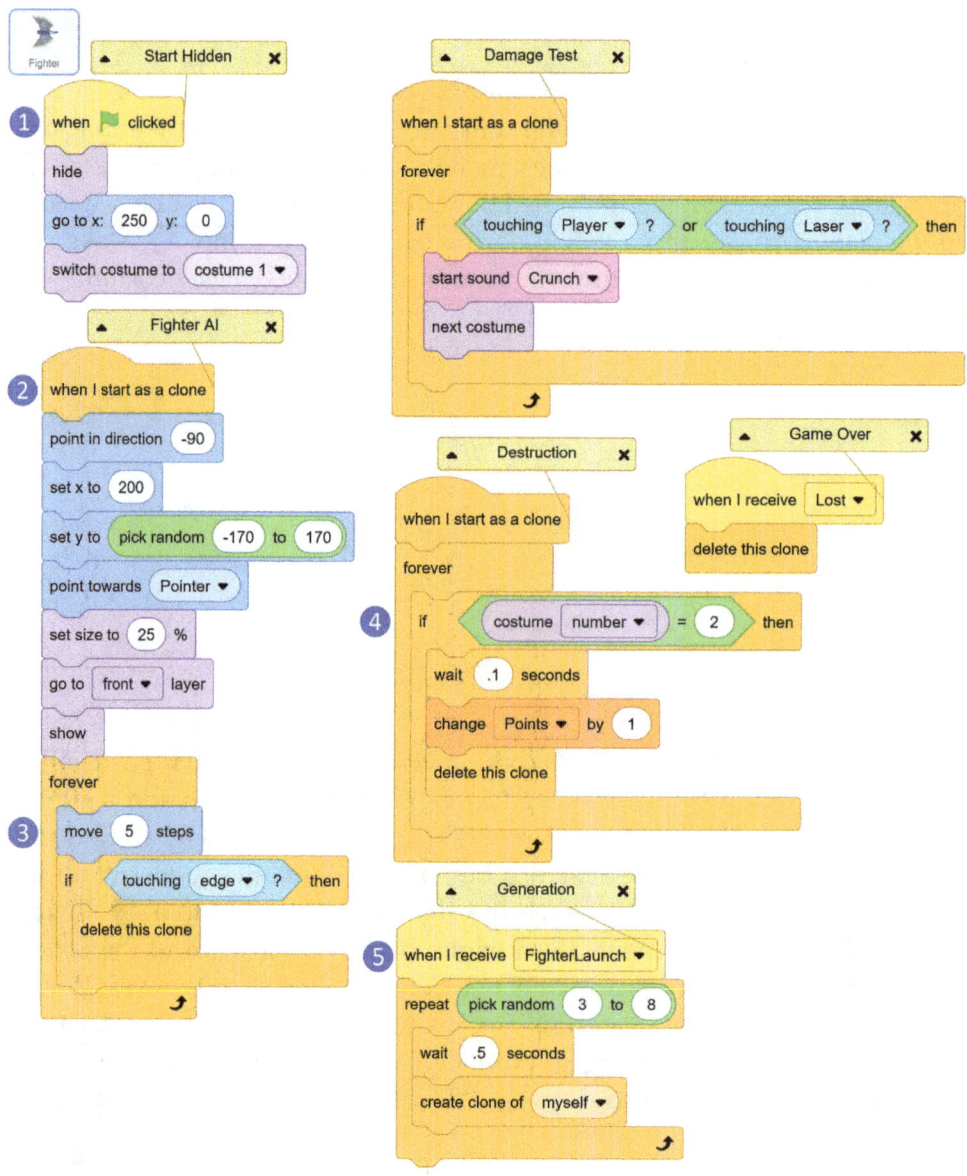

Advanced Project 4: Scrolling Shooter ♦ 133

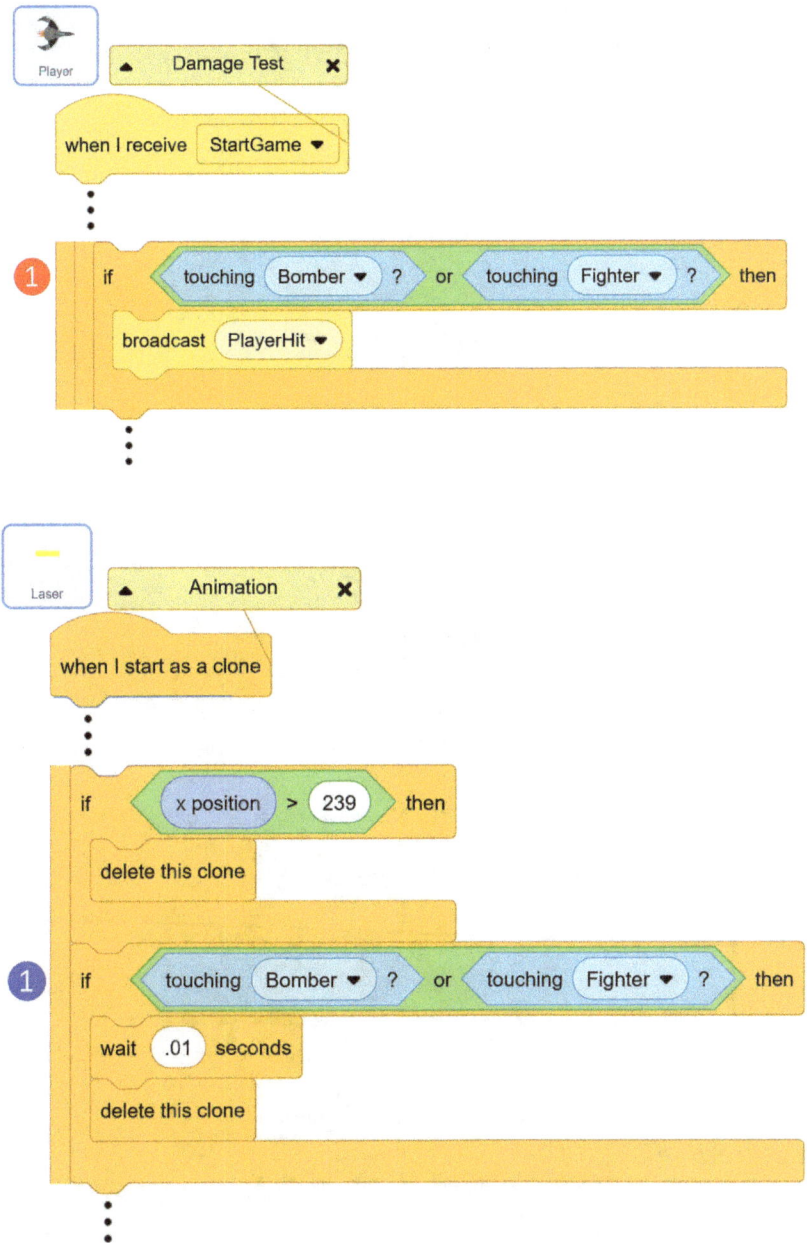

In the *stage* code, add a ① •*"FighterLaunch"* event to the 3, 5, 6, and 7 •*"WaveType"* •*Ifs*. We've now got a nimble *fighter* enemy, but what if they were even more dangerous?

134 ◆ Advanced Project 4: Scrolling Shooter

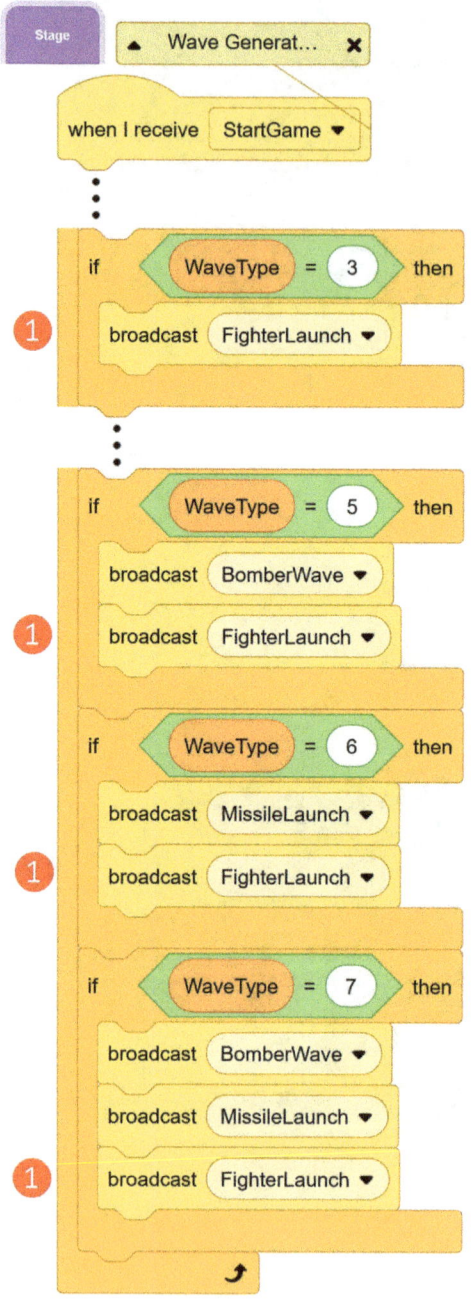

Step 10: Now with More Lasers!

We're going to make things a little more complex by allowing the *fighter* enemy to shoot lasers of their own. Start by adding a new blank sprite, "*EnemyLaser*", and for a costume simply make a small red rectangle so it's distinct from the *player's* yellow *laser*. If you want something fancier, use gradients of red to transparency with two adjacent rectangles vertically flipped to make a glowing red laser.

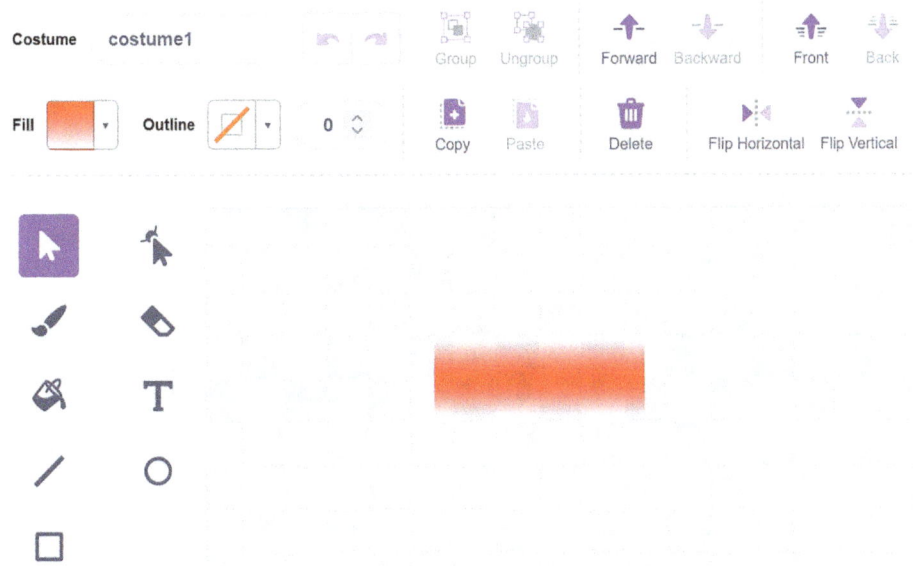

For its code, ① start it with a •[Hide], and in a ② •"*Lost*" event, we • [Delete This Clone]. Most of the code is in a ③ •[When I Start As A Clone] event. We need the *EnemyLaser* to start at the location of the enemy that shot it; however, if we refer to the *fighter* sprite, it will use the original sprite and not the clone doing the shooting. So we need to create three new hidden variables: •"*EnemyLaserX*", •"*EnemyLaserY*", and •"*EnemyLaserDir*". Set the location ④ to •[Go To X: •("*EnemyLaserX*") Y: •("*EnemyLaserY*"), •[Point In Direction •("*EnemyLaserDir*")], and •[Show]. With these we'll be able to get around the original/clone inheritance/reference issues. In a •[Forever] loop, add in movement and two conditions to delete the clone, ⑤ when it touches the edge or ⑥ when it's touching the *player*. You'll notice we use a • [Wait] when colliding with the *player*. This will help make the hit more visible while also ensuring that the *player* can react to colliding with the *EnemyLaser* before the *EnemyLaser* disappears.

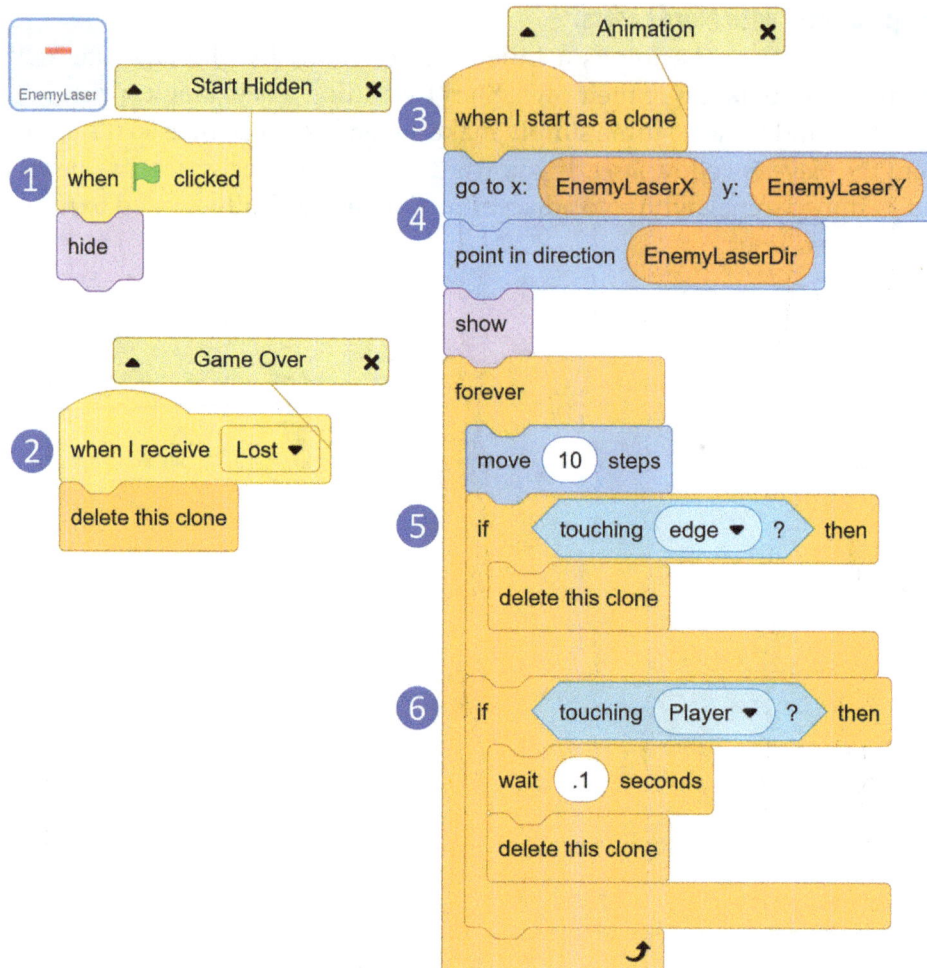

Next, update the *fighter* to allow it to shoot. In the *//Fighter AI* stack, add an ① •[If •<•(Pick Random (*1*) to (*100*)) < (*2*)> Then]. This will basically roll a dice, and only on a 1-in-100 chance will the *fighter* shoot. It makes it random as to when or how often they shoot but keeps it rare enough. It might seem like very low odds, but keep in mind it will roll the dice every step, so expect a shot every 3 seconds from each *fighter*. In this •*If*, ② set our three new variables to the current X position, Y position, and direction and then ③ •[Create Clone of [*EnemyLaser*]]. Because we set these variables from the clone, they inherit the clone's properties, but trying to use ([X *Position*] of [*Fighter*]) would just use the original fighter's properties, not the specific clone's, so we have to use variables to pass the data correctly.

Advanced Project 4: Scrolling Shooter ◆ 137

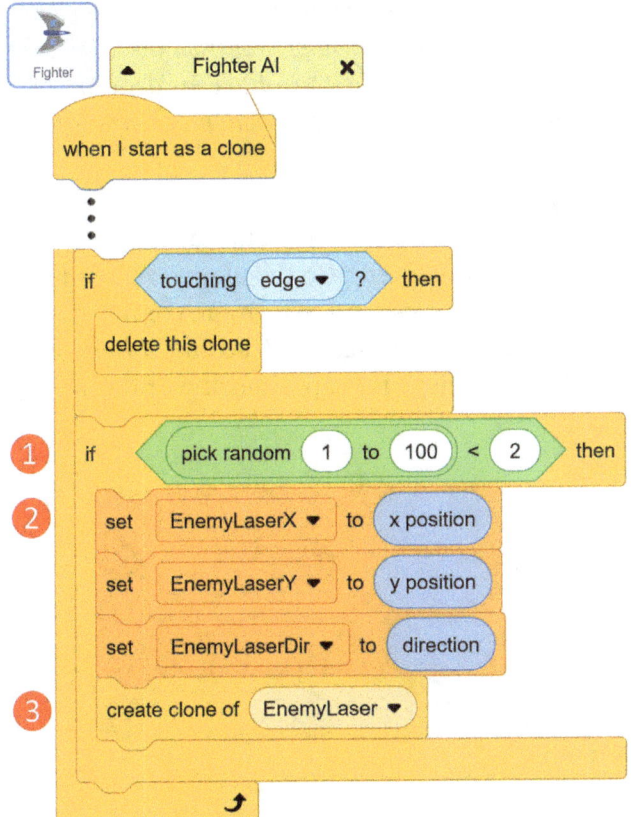

With the *EnemyLaser* created, update the *player* to collide with it. In the *player's* //Damage Test stack ①, duplicate a test for •<Touching [Enemy-Laser]?> or add it to an existing one with an •<<condition> or •<Touching (EnemyLaser)>>. Don't update *laser* since we don't want the *player* to be able to shoot a laser with a laser.

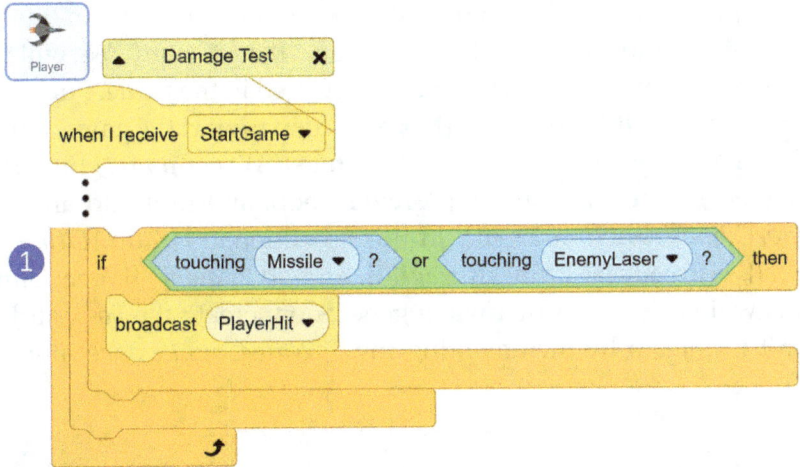

With your now-advanced understanding of working with clones, you might want to revisit the Fireworks Display project from Book 1: Beginner and try to use clones. Conversely, you might want to consider some of the fireworks effects in that project as a way to spice up the special effects for this game!

Step 11: The Boss Approaches

Our next task is to add the final *boss* for our game, a bigger, badder enemy that will be the last, biggest challenge to the player. When defeated, the player wins the game. In games, bosses tend to be more complex by having different forms or modes, doing more things, and lasting longer while keeping the player guessing. This means a lot more work to program than a standard enemy.

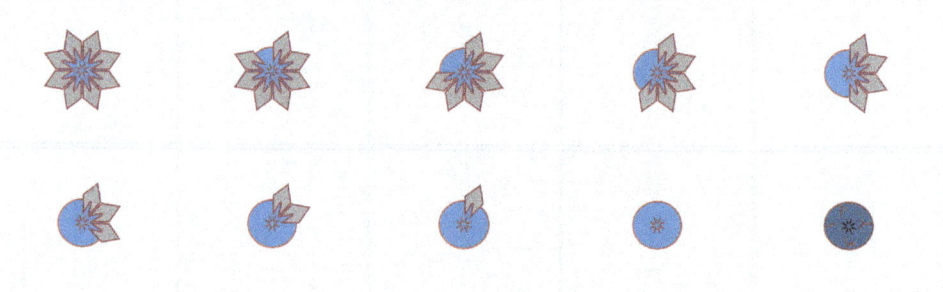

Start by making a blank sprite *"boss"*. In the Costumes tab, we're going to need a number of different costumes to display the gradual destruction of this very difficult enemy. I've used ten different costumes, made a base object with both a functioning and destroyed version of it. Then, I added eight armour sections to it. It starts at the fully armoured version, then each costume removes one piece of the armour until getting to the base object at costume 9, and then the destroyed version at 10.

First, spawn the *boss*. To do this, start with the *stage* code. In our //Wave Generation stack, add a new hidden variable, •*"WaveCount"*. We will use this variable to track how far the player has progressed in the game. Put a ❶ •[Set ["WaveCount"] To (0)] above the •[Forever] loop to start the game, and below the •[Set "WaveType"], add in a ❷ •[Change ["WaveCount"] by (1)]. Then, add another ❸ •"GameStart"/•[Forever] loop, and in it add an ❹ •[If • <•("WaveCount") = (20)> Then]. In this conditional, •[Stop [Other Scripts In Sprite]], •[Broadcast ["BossReveal"]], and •[Change ["WaveCount"] by (1)]. This will stop the normal wave generation, create the *boss*, and ensure it doesn't run again by incrementing the •*"WaveCount"* once more. While

Advanced Project 4: Scrolling Shooter ◆ 139

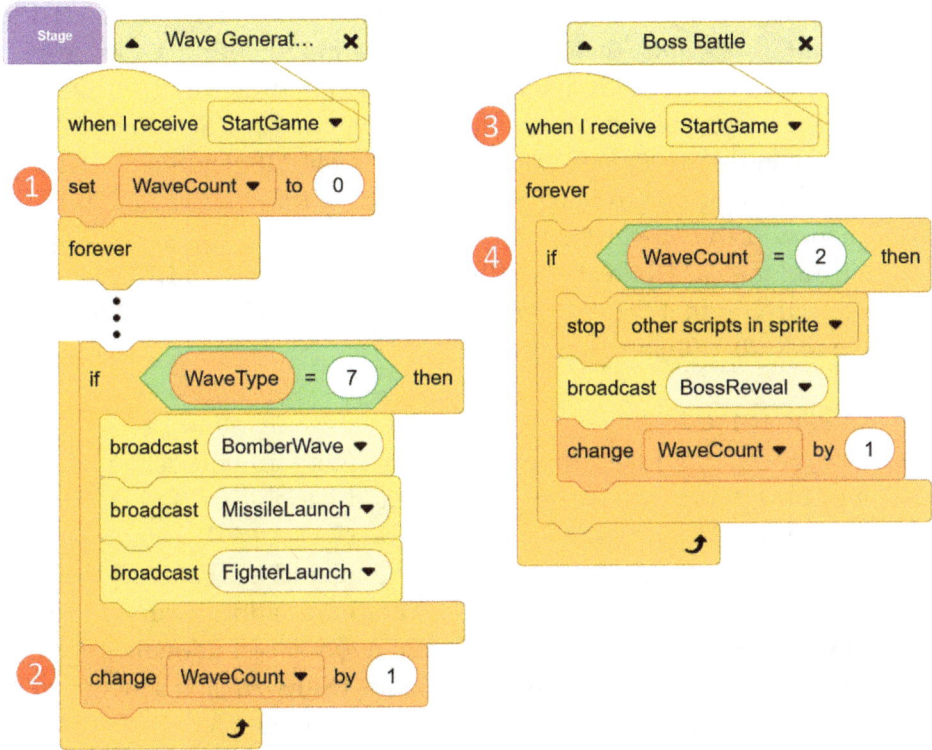

testing, you can set the ④ •[If •<•("*WaveCount*") = (2)> **Then**] to jump to the *boss* right away so it doesn't take forever to see if the boss is working right. Just remember to change it back when you're done testing!

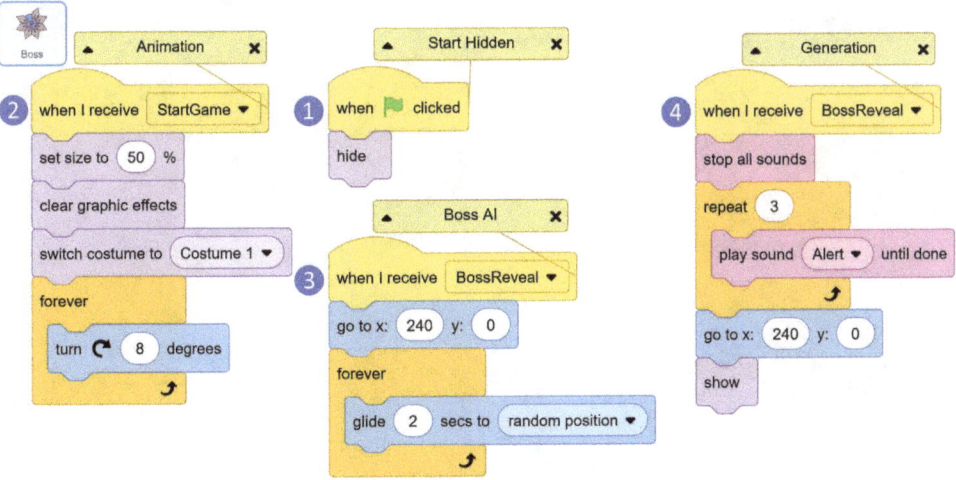

Next, start coding the *boss*. As usual, we start the game ① with •**[Hide]**. We'll add a ② •**"GameStart"** event to handle graphics and set its costume to 1. Animate the *boss* with a simple •**[Forever]** •**[Turn Right (8) Degrees]** tucked at the bottom of the //*Animation* stack. Start a ③ •**"BossReveal"** //*Boss AI stack* to handle movement, and have it at **X240 Y0**, then go a •**[Forever]** loop and add a single •**[Glide (2) Secs to [Random Position]]** as a stand-in for now. Then, add the //*Generation* ④ •**"BossReveal"** event. Here, •**[Stop [All Sounds]]** and then a •**[Repeat (3)]** •**[Play Sound [Alert] Until Done]**, giving a dramatic warning that the *boss* is about to approach. After the •**[Repeat]**, we can •**[Go to X:(0) Y:(0)]** and •**[Show]**.

Step 12: Fighting the Boss

Now that we have our *boss* appearing and moving around, let's make sure we can fight it. Start by adding a new hidden variable, •**"BossHP"**. In the *boss* add another ① •**"BossReveal"** event for a //*Damage Test* stack and start it with a •**[Set ["BossHP"] to (100)]**. Under that, add a •**[Forever]** loop, and in it, test ② •**[If** •**<Touching (Laser)?> Then]** and add a **Crunch** SFX and •**[Change ["BossHP"] by (-1)]**. Under it, add another conditional, ③ •**[If** •**< ("BossHP") < (0) Then]**, and then •**[Hide]** and •**[Stop [All]]**. It's inelegant but works as a placeholder to know we can defeat the boss.

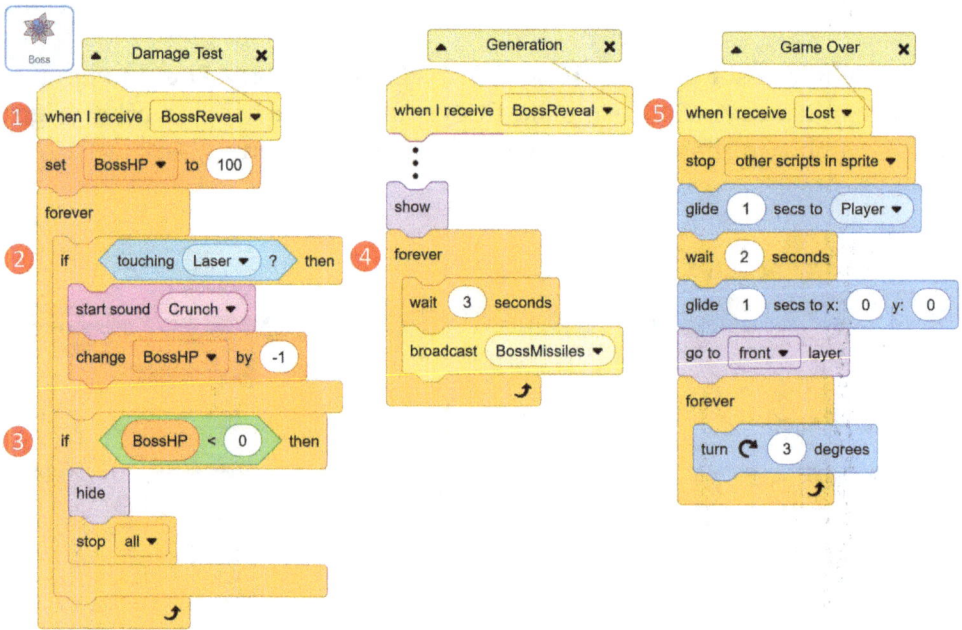

Then, in the //*Generation* stack, add at the bottom a ④ •**[Forever]** •**[Wait (3) Seconds]** •**[Broadcast ["*BossMissiles*"]]**. Add a ⑤ •**"*Lost*"** event. This will •**[Stop [*Other Scripts In Sprite*]]**, •**[Glide (0.5) Secs To [*Player*]]**, •**[Wait (2) Seconds]**, then •**[Glide (1) Secs To X: (0) Y: (0)]**, •**[Go To [*Front*] Layer]**, and then •**[Forever]** •**[Turn Right (3) Degrees]**. This is a cinematic sequence where the *boss* will crush the *player*, then centre on the screen and rotate slowly. We'll add it to our Game Over sequence later, but this should be a good start.

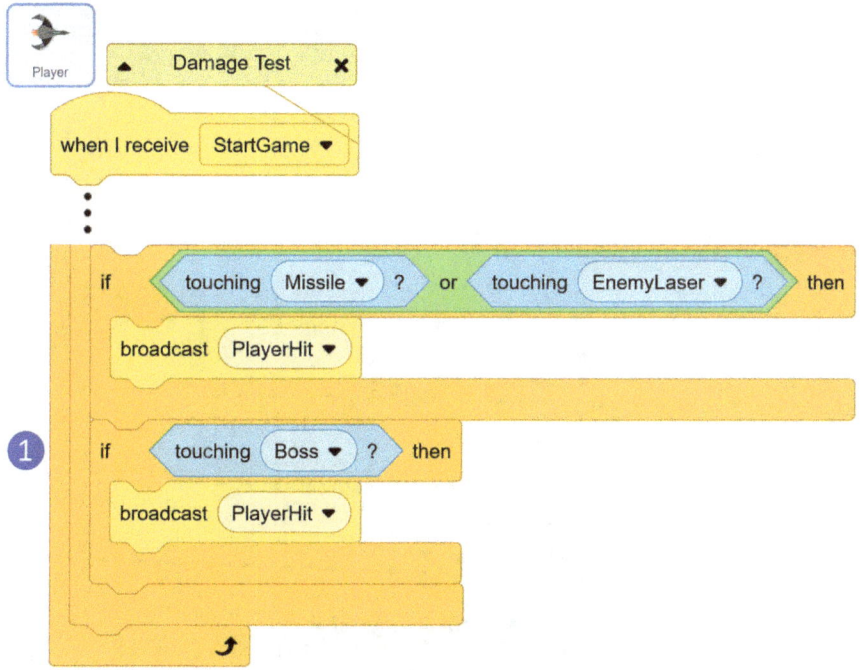

Next, we need to update our *player*, *laser*, and *missile* to interact with the *boss*. In our *player*, add a ① •**[If •<Touching [*Boss*]?> Then]** test to broadcast •"*PlayerHit*" in our collision-testing stack. In our *laser*, ① duplicate any of the •**Touching** enemy •**Ifs** and switch it to the *boss* so our *laser* self-destructs after hitting the *boss* instead of continuing to sail through them. Lastly, in our *missile*, add a new ① •"*BossMissiles*" event. In it call •**[Go To [*Boss*]]** and •**[Create Clone of [*Myself*]]**. This way, the *boss* is always launching a stream of *missiles* at the *player* to keep them on their toes.

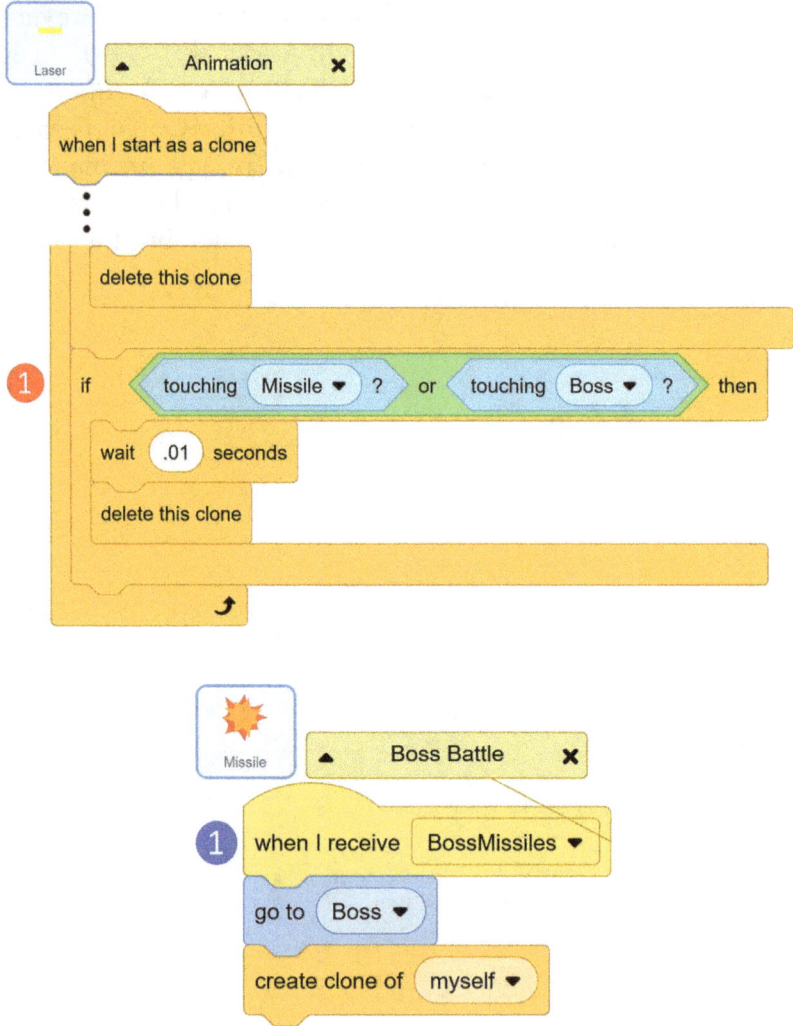

Step 13: Boss Movement Patterns

Our next step will be to get the *boss* moving around the screen to be a bigger threat to the *player*. Start by adding a new hidden variable, •*"MovePattern"*. In our //*Generation* stack in the *boss* code, start with a ① •[Set ["MovePattern"] to (0)], then add and set it to •(Pick Random (-2) to (4))]. This will switch up the *boss's* tactics every three seconds by randomly selecting a new movement pattern.

To switch up our //*Boss AI* stack, we'll add a bunch of •[If •<•("MovePattern") = (#) > Then] comparison conditionals. Start with ② •*"MovePattern"*<1. Move our •[Glide (2) Second To [Random Position]] to inside this • If. You can copy along the other •*"MovePattern"* If's provided. ③ Number 1 makes the *boss* move back to the right-side centre. ④ Number 2 makes them

Advanced Project 4: Scrolling Shooter ◆ 143

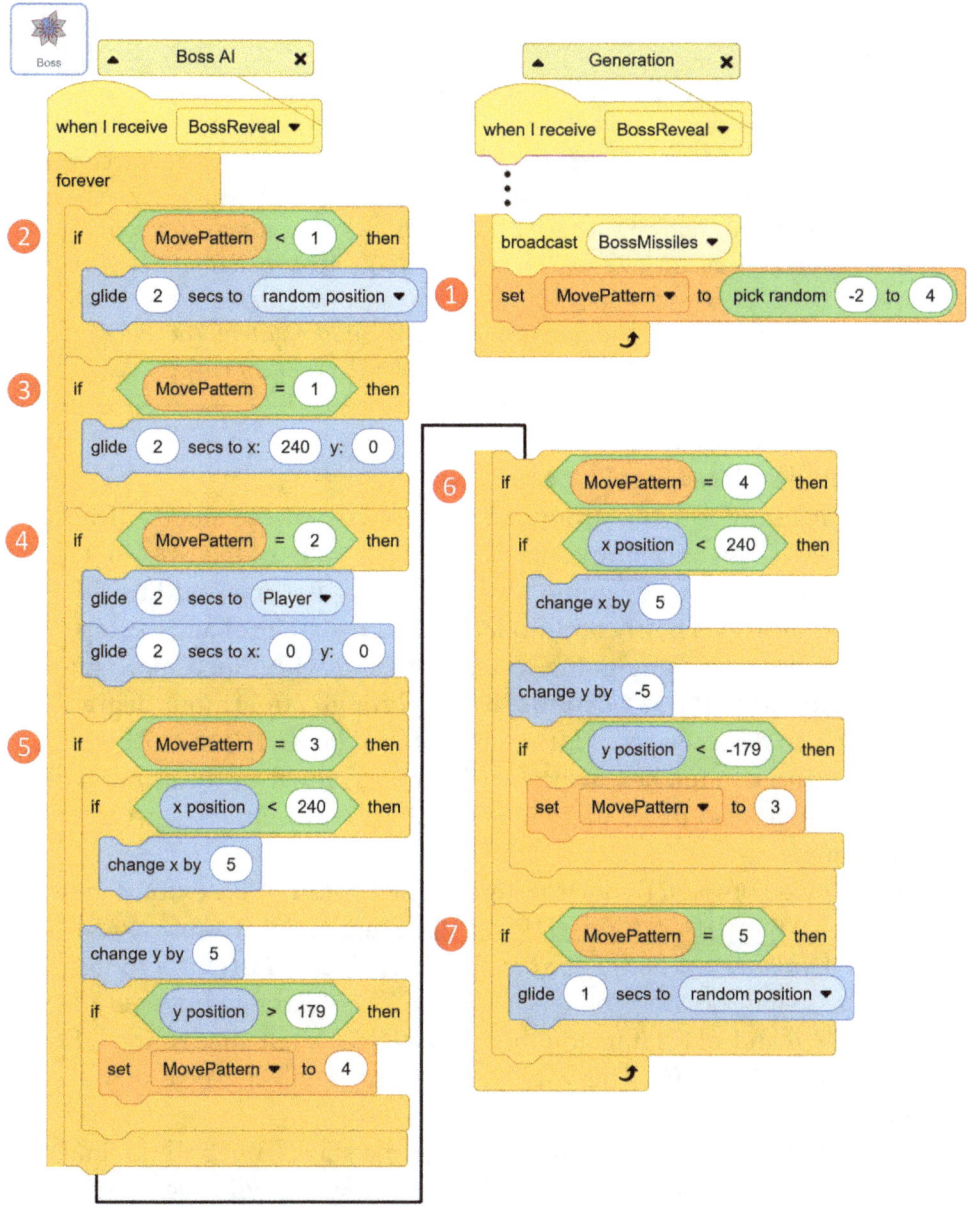

move to the player, then the centre of the screen. ❺ Number 3 makes them move to the top right corner. ❻ Number 4 makes them move to the bottom right corner. Lastly, ❼ Number 5 makes them move to a random position even faster than usual. You'll notice that the 3 and 4 movement patterns end in switching the movement pattern, so the *boss* will go up and down along the right side for 3 seconds. You may have noticed that pattern 5 couldn't happen with the ●**(Pick Random)** values we're using.

> **Smarter AI**
>
> *In this project we're just using a random choice for what movement pattern to do, but AI systems can be a lot more devious. In most commercial games, AI will have conditional, or analysis-based decision-making, instead of just random chance or set patterns. Try the system as it is for now, but if you really want to master game development, try making an enemy that senses the player or game conditions to make its choices. Maybe it determines the player's distance to determine its weapon use. Maybe it determines the player health to be more or less aggressive, either making a set choice or even just changing the odds of a random system. There are lots of ways to make decisions, and enemy AI is a great way to explore them in code!*

Step 14: Boss Health and Boss Phases

With the *boss* now moving around, we can focus on the stages of the *boss* fight. We made a lot of costumes for the damage states of the *boss*, so it's time to put them to use. Start by going to our *//Damage Test* stack and its ❶ •[If •<•("BossHP") < (0)> Then] conditional. Delete the •[Hide] and •[Stop [All]] code blocks because we're going to replace them with much more interesting code. Add in a •[Next Costume] to run through our armour degradation cycle, and also reset the *boss's* HP with a •[Set ["BossHP"] to (25)]. This will heal the *boss* back to full health after each piece is blown off the armour. However, this would also make them invulnerable, because there's no condition to end things!

Before we completely destroy the *boss*, let's add in an extra dangerous mode. Above the •[If •<•("BossHP") < (0)) Then], add in a new ❷ •[If •<• (Costume [Number]) = (9)> Then] and a •[Set ["MovePattern"] to (5)]. This will ensure that if the *boss* is on its last phase, it acts extra-aggressively. Duplicate this ❸ *If* and place it inside the •"BossHP"<0, but change the •[Set] to •[Set ["BossHP"] to (50)]. This way, the final phase will also be much tougher than the previous ones.

Now we have the *boss* progress through the stages, but still no final end. Next, let's add our Game Over condition. Below the *//Damage Test* stack's •[If •<•(Costume [Number]) = (9)> Then], add in a new ❹ •[If •<•(Costume [Number]) = (10)> Then]. This will trigger when the boss finally takes the last point of damage in their last phase. Inside it, •[Stop [Other Scripts] in Sprite, •[Go to [Back] Layer] then ❺ run a •[Repeat (5)] with a •[Change [Ghost] Effect by (20)], •[Wait (0.1) Seconds], •[Start Sound [Connect]], and then a ❻ •[Repeat (50)] with a •[Change ["Points"] by (1)]. This will create an animation fading out the *boss* while racking up a bunch of •"Points" for the *player*. Below the •[Repeats], but still inside the •[If •<•(Costume[Number]]) = (10)> Then], add a ❼ •[Hide] and •[Stop [All]].

Advanced Project 4: Scrolling Shooter ◆ 145

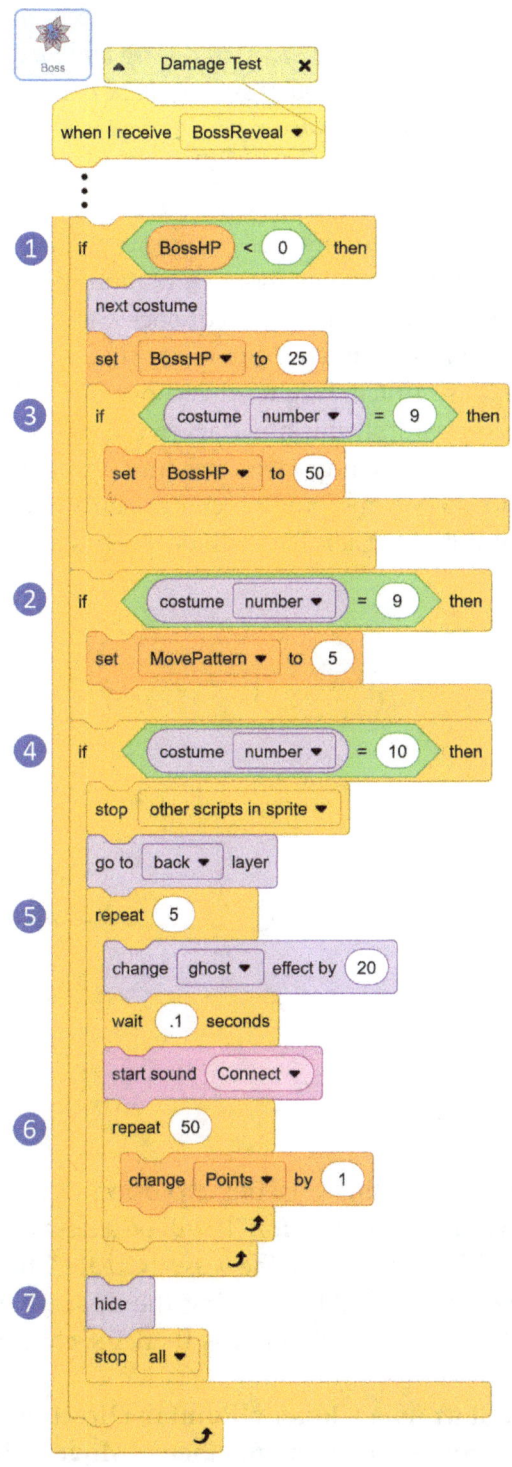

We didn't have room to add in power-ups or collectables in this project, but we've got examples of both in our Big Map Racing and Snowball Fight game projects in Book 2: Intermediate, so be sure to check those out for ideas about how you could expand on this game!

Step 15: Victory!

We've almost finished with the *boss*. Let's make sure our victory is as memorable as it deserves by adding in some more animations! Start with an explosion appropriate for the *boss* and a *"Victory"* sequence. Start adding a blank sprite, *"Explosion"*. This animation is made from a star shape and a number of small circles. Start by making a nice, bright spread-out but still touching version, then duplicate it and spread, shrink, and darken the components, repeating the process a few times until the components are tiny. Then take the original middle version and duplicate it, moving the components closer together, and do so again to create an initial cluster. This will be our basic explosion for the *boss*.

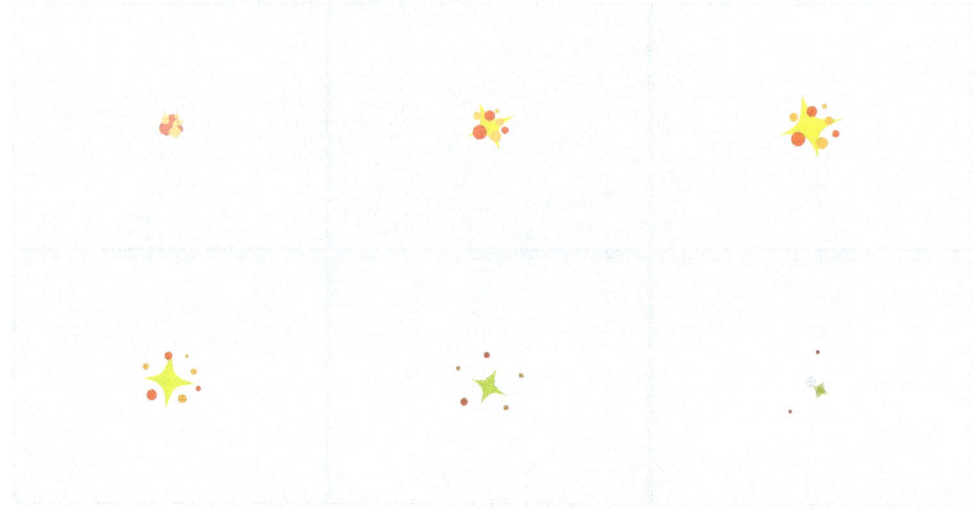

With code, we'll make this an epic explosion. ① Start with ●[Hide]. ② ●[When I Receive [*"BossExplosion"*]] will be a new event to generate the explosion sequence. Have it move to the *boss*, ●[Create Clone of [*Myself*], and ●[Start Sound [*Space Noise*]]. After adding the **Space Noise** clip, you may want to trim it to be shorter – I like it less than 1 second long, with the last portion faded out.

In ③ ●[When I Start As A Clone], ●[Point in Direction ●(Pick Random(1) to (360))] to have it rotate to any angle, then ●[Move ●(Pick Random(6) to (40)) Steps] so it offsets at random so that it'll spread out to cover the large size of the *boss*. Next, ●[Switch Costume] to the first costume and ●[Show].

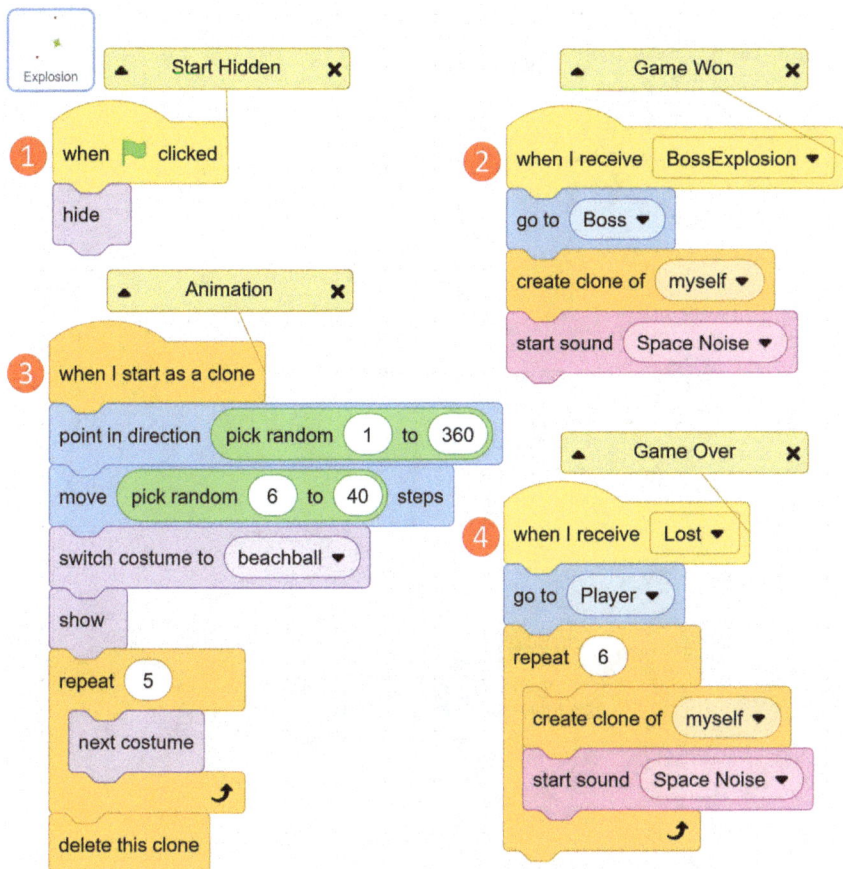

Then, use a ●[Repeat (5)] to ●[Next Costume] our way through our sequence of six images. After, we ●[Delete This Clone]. We've now got a great, dramatic explosion fitting of the *boss*.

Lastly, in a ❹ ●*"Lost"* event, have the *explosion* ●[Go To [*Player*]] ●[Repeat (6)] ●[Create Clone of [*Myself*]] ●[Start Sound [*Space Noise*]] so that if the *player* loses the game, they'll also suffer an epic explosion ending their game.

Go back to the *boss's* code to call some ●*"BossExplosions"*. Add one to the ❶ ●[If ●<●("BossHP") < (0)> Then], as well as both ❷ in and ❸ after the ●[Repeat (5)]. Replace the ●[Stop [*All*]] with a new ❹ ●*"Victory"* event and a ●[Stop [*This Script*]]. This will give a big explosion when each piece of armour is removed, as well as a sequence of explosions when the *boss* is totally defeated. Next, we'll need to update a few other sprites. In the *missile*, add a ●*"BossExplosion"* event and have them ●*"Explode"* so they don't risk killing the *player* after they've defeated the *boss*.

148 ◆ Advanced Project 4: Scrolling Shooter

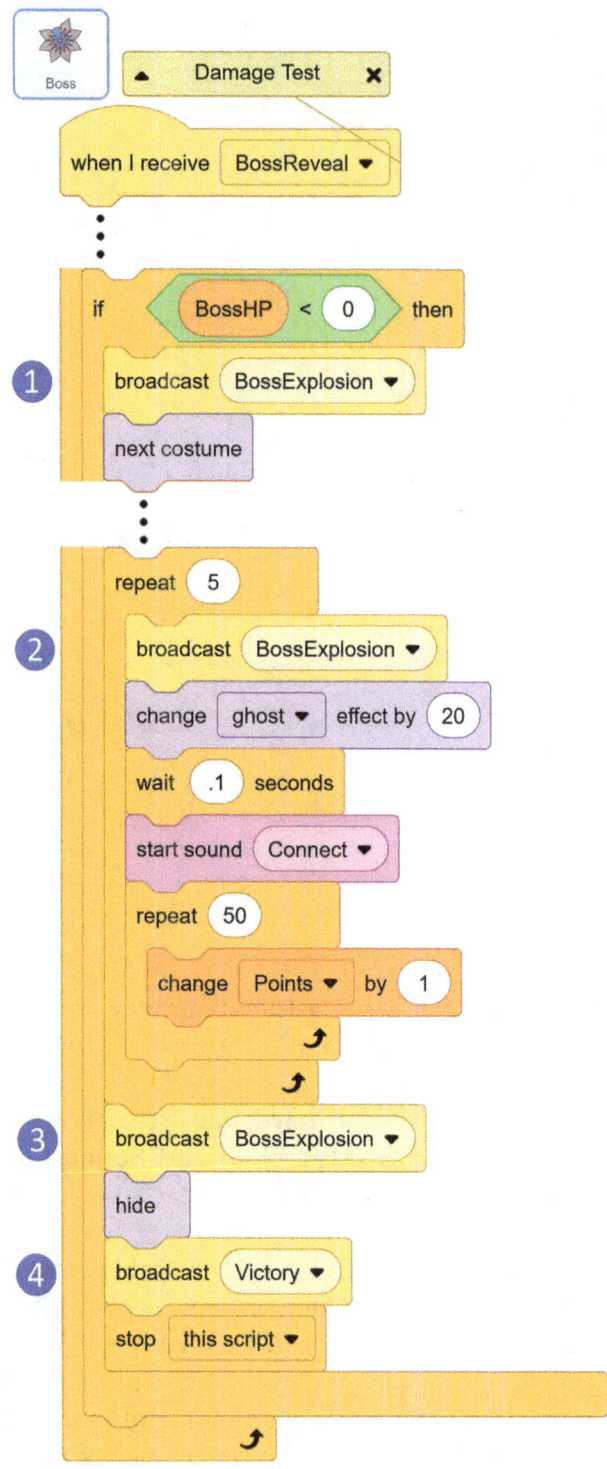

Advanced Project 4: Scrolling Shooter ◆ 149

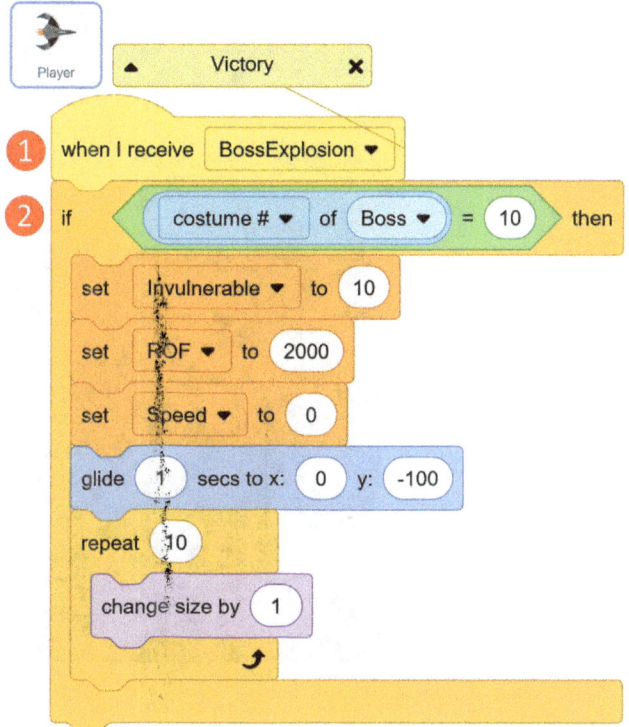

Player will need a ①•*"BossExplosion"* event too. We'll test that it's the final phase of the *boss* with a ②•[If •< •([Costume Number]of [Boss]) = (10)> Then] and v[Set ["*Invulnerable*"] to (10)], [Set ["*ROF*"] to (2000)], and •[Set ["*Speed*"] to (0)]. These will render the player essentially non-functional. Then, •[Glide (1) Second to X: (0) Y: (-100)] and •[Repeat (10)] •[Change Size By (1)]. This will act as a little victory animation, having them flying to the middle and seemingly toward the player with the size change.

In the *SpaceBG*, create another costume. Copy the Loss costume and turn it into a Victory costume. Add in a "Victory!" text at the top and a little story text, like, "You defeated the invading alien fleet". We want to ensure there's room for the *player* at X: (0), Y: (-100) without them covering up the text. In the code, add a ①•*"Victory"* event, where we'll •[Stop [*All Sounds*]], •[Stop [*Other Scripts in Sprite*]], •[Go to X: (0) Y: (0)], and •[Switch Costume to ["*Victory*"]]. This will stop the scrolling effect and give us our "*Victory*" background. Finally, we can •[Play Sound [*Trap Beat*] Until Done] as celebratory music.

150 ◆ Advanced Project 4: Scrolling Shooter

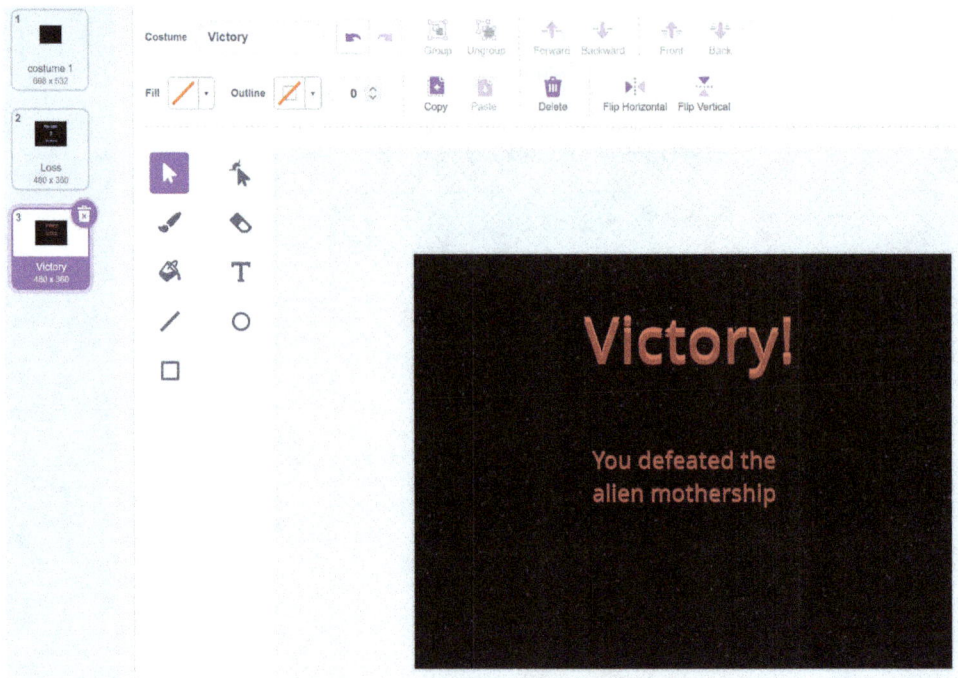

Step 16: Main Menu

Now that the gameplay is settled, we can add the *menu*. Some of you may be wondering why we always wait until the end to add a menu. It's so we don't have to go through the delay of a menu when testing a game. It's always nice to be able to jump right to what you need to test, so skipping the menu until after you get your game in order is a nice way to save time.

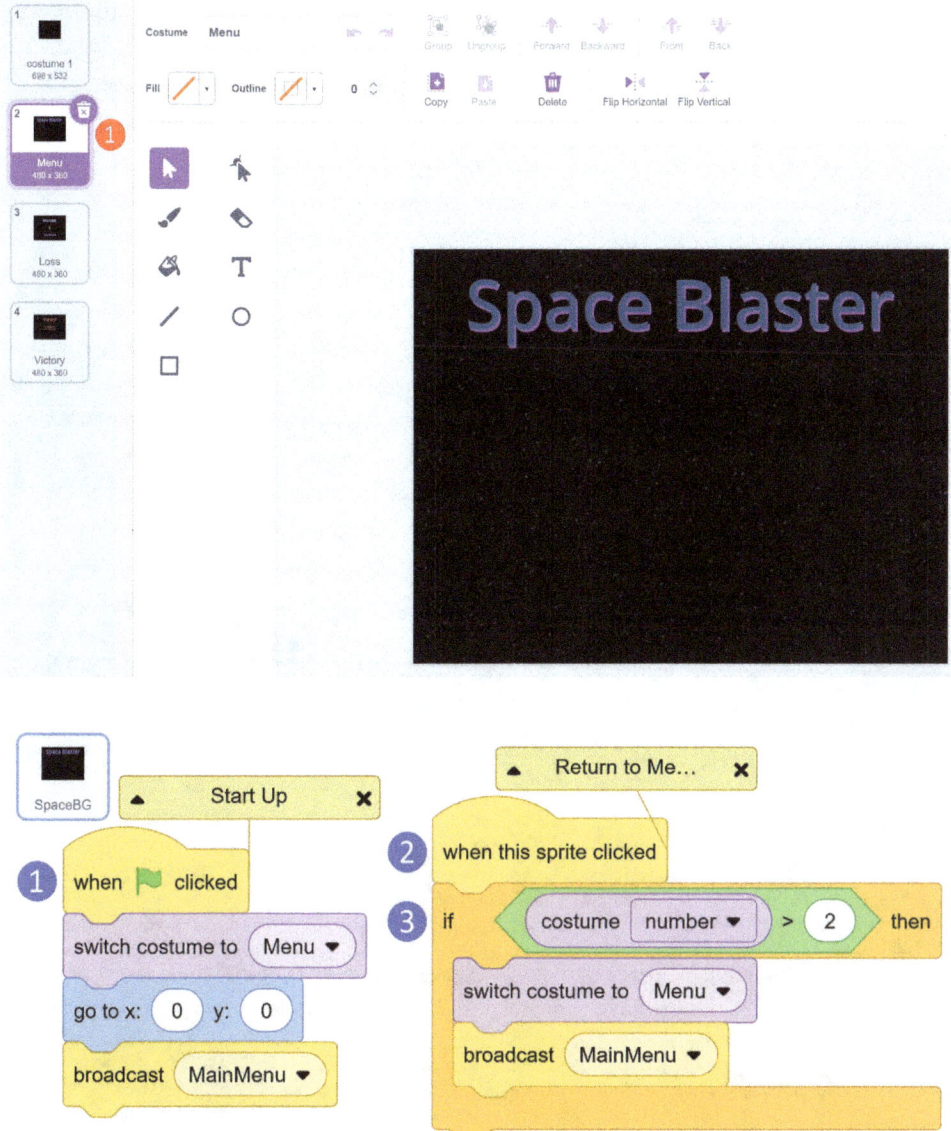

Start by adding another costume to our *SpaceBG* to serve as the main menu BG. You can copy the "Loss"/"Victory" costume and just switch up the text, naming the game with a title and calling this costume "Menu". Reorder the costumes ❶ so the "Menu" is #2, with "Loss" and "Victory" after it. Next, we'll add some code. ❶ •**[When ▷ Clicked]**, we need to •**[Switch Costume to [*Menu*]]** and •**[Go To X: (0) Y: (0)]**, and •**[Broadcast ["*MainMenu*"]]**. This will ensure the game starts on the menu but also give an event to correct other objects to a menu state. Add in a ❷ •**[When This Sprite Clicked]** and a conditional ❸ •**[If •<•(Costume Number) > (2)> Then]** to trigger only when it's on the "Loss" or "Victory" costumes. Inside it, place a •**[Switch Costume to [*Menu*]]** and •**[Broadcast ("*MainMenu*")]**.

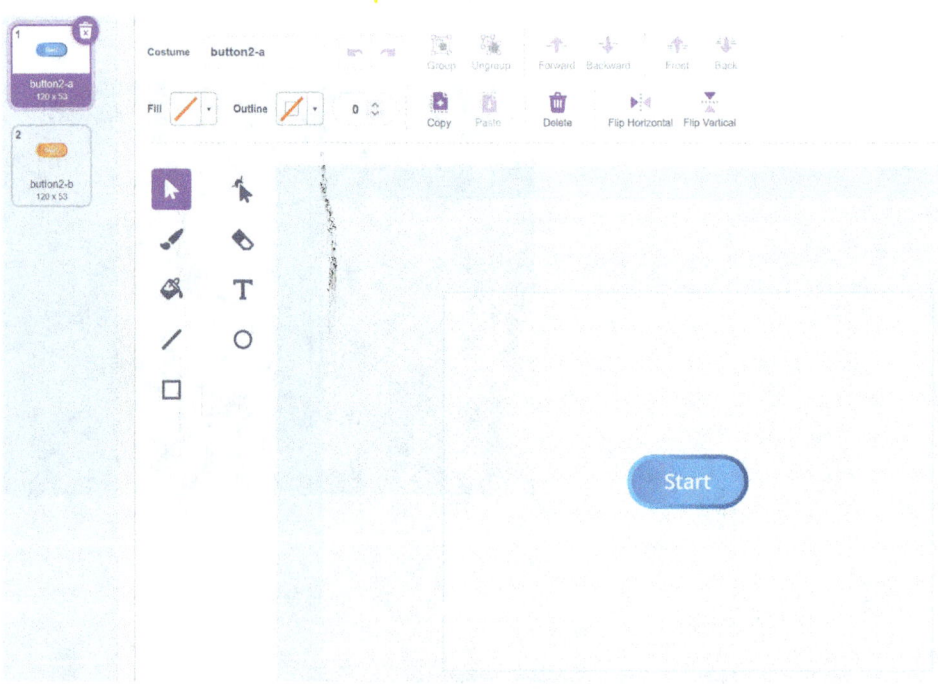

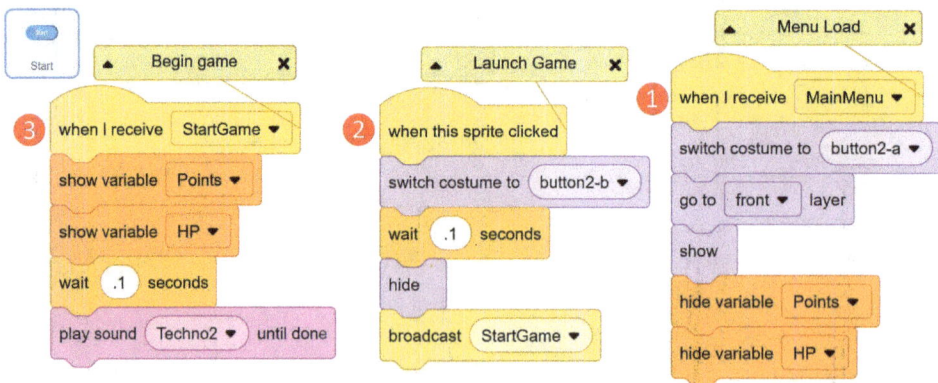

Next, add a *Start* button for our main menu to launch the game. You can add a **Button 2** from the Sprite Library, rename it "*Start*", and add "Start" text to the **Costumes**. In the **code,** we'll need a ❶ •*"MainMenu"* event to •[**Switch Costume to [***Button2-A***]**], •[**Go To [***Front***] Layer**], •[**Show**], and •[**Hide Variable**] for •*"Points"* and •*"HP"*. We'll also need a ❷ •[**When This Sprite Clicked**] to •[**Switch Costume to [***Button2-A2***]** (so it shows it being clicked), •[**Wait (0.1) Seconds**], •[**Hide**], and •[**Broadcast [***"GameStart"***]**]. Add a ❸ •*"GameStart"* event and •[**Show Variables**] for •*"Points"* and •*"HP"*. Optionally, you can add a •[**Wait (0.1) Seconds**] and •[**Play Sound [***Techno2***] Until Done**] to give some hype music to start the game with.

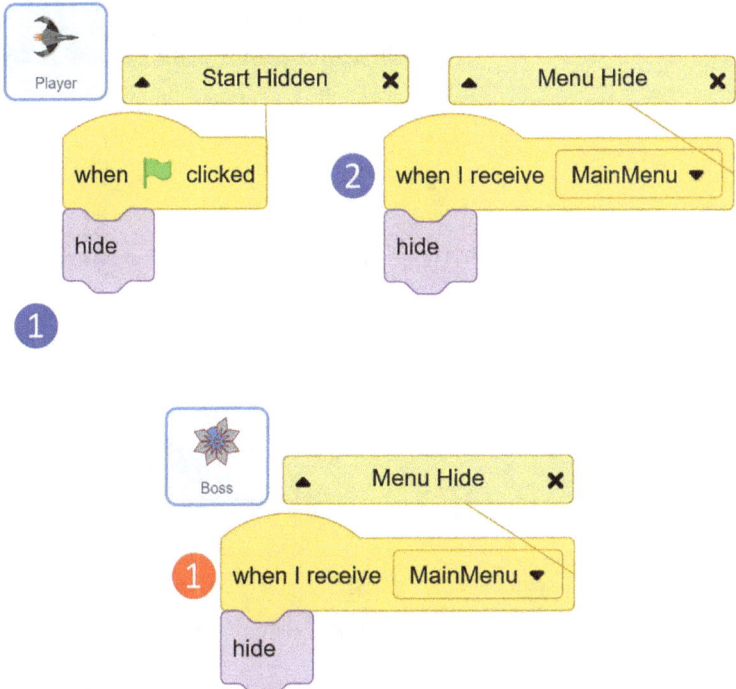

Our *player* will need updating now. ❶ Remove the •[**Broadcast [***"Game-Start"***]**] from our *player* code since it's now handled by the *Start* button. Also, add in a ❷ •*"MainMenu"* event to •[**Hide**]. The *boss* will also need a ❶ •*"MainMenu"* event to •[**Hide**].

Step 17: Leader Boards
With our random generation of enemies and the variance of play, it's fun and engaging to let players compete for high scores. Let's add a score-tracking system using Scratch's •**Cloud Variables**. This will allow us to save the results of play not just for a single player but for players around the world. To do this, we'll need to add score tracking and display.

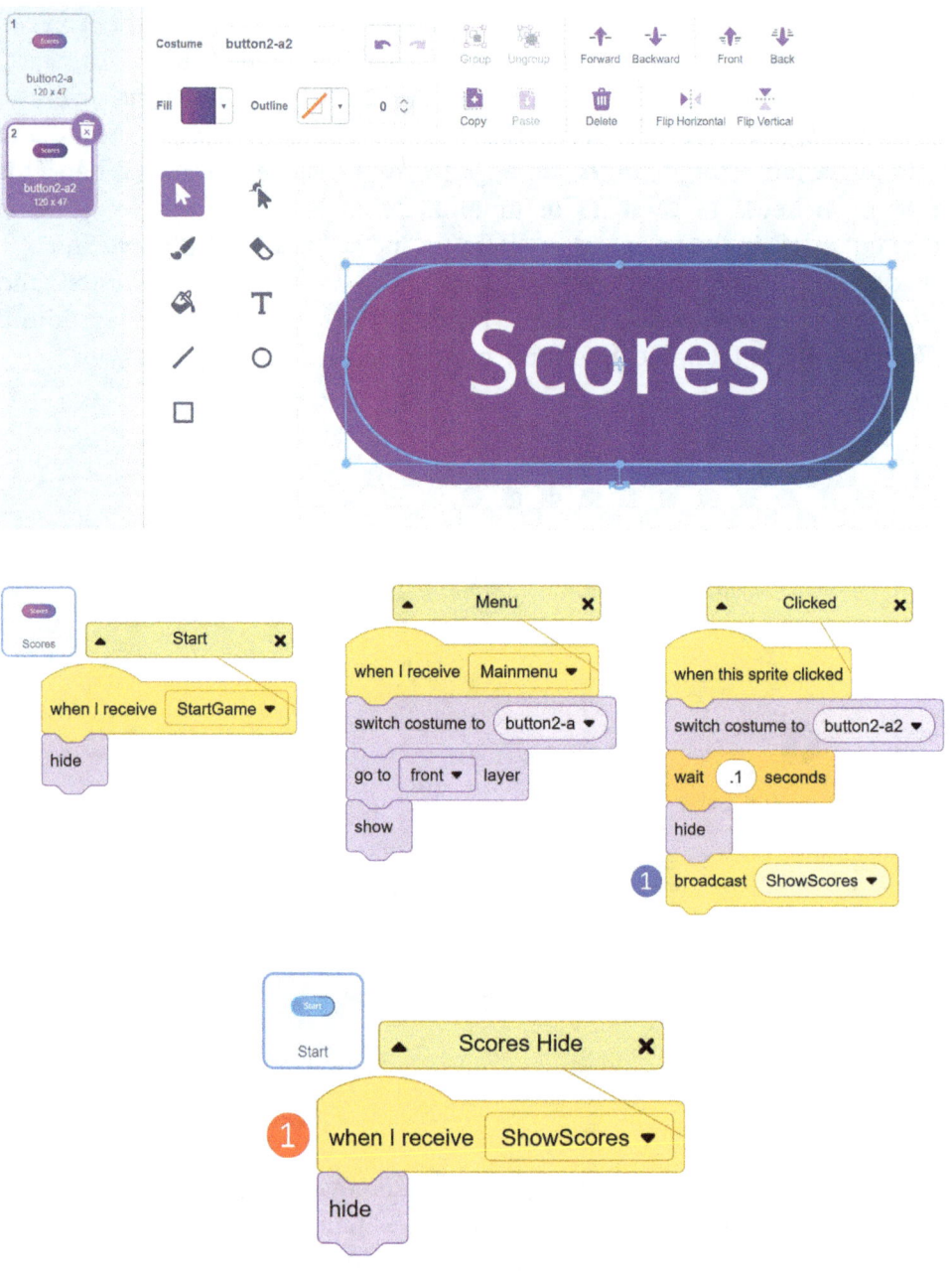

Start by adding a *Scores* button to display the scores. Duplicate the *Start* button and tweak the graphics to a different colour and text. Rename it **"Scores"**. This button will take us to a view of the global leader board for top scores. We'll need to change the •**[Broadcast]** to call a new ❶ •*"ShowScores"* event. We won't need the •*"Start"* game to do anything but •**[Hide]**. Edit our *Start* button as well, adding a ❶ •*"ShowScores"* event to •**[Hide]** the sprite.

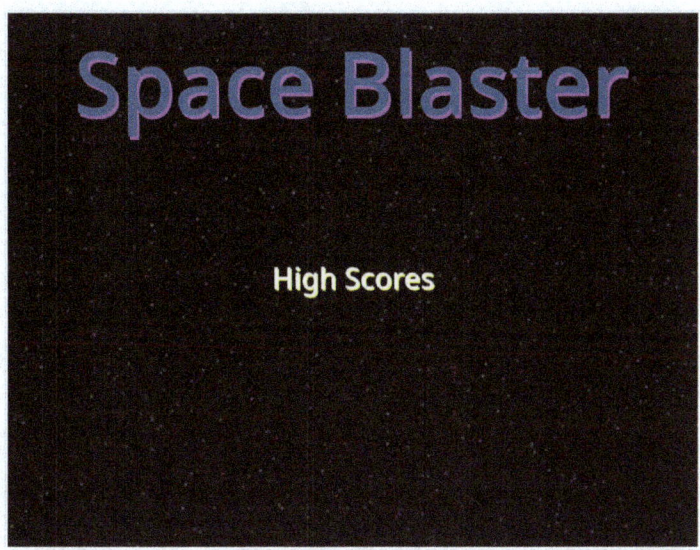

Our main task will be creating a new blank *"Leaderboard"* sprite. Copy and paste our Menu costume from *SpaceBG* to this new sprite, but add an extra line of text, *"High Scores"*, just below the title. In the code, we'll need to do a bit of repetitive work to handle score-keeping. Start with a ① •*"Lost"* and •*"Victory"* event both with •[Broadcast [*"SetScores"*]]. Add a ② •[When This Sprite Clicked] to call •*"HideScores"* and •*"MainMenu"*. Start the game using ③ •[When ▷ Clicked] •[Broadcast [*"HideScores"*]] and •[Hide]. Before we go further, we'll need to create five new hidden variables, each as a •Cloud Variable (mark the checkbox when creating a variable). These variables are •*"1st Place"* through •*"5th Place"*, allowing us to track the top 5 scores in our game globally.

In ④ •*"ShowScores"*, •[Show] and a •[Show Variable [variable]] for each of the five score variables. In ⑤ •*"HideScores"*, •[Hide], and a •[Hide Variable] for each of the five score variables.

We will need a series of nested ⑥ •If/Else conditionals to test what the player's •*"Points"* were versus the global high scores and adjust the global high scores if necessary, adding their score at the appropriate level. This can be a bit repetitive as we need to test each position and bounce the scores down the list according to any given level achieved. Start with the lowest position. Make a conditional •[If •< •(*"Points"*) > •(*"5th Place"*)> Then] and •[Set [*"5th Place"*] to •(*"Points"*)]. The player's score will simply bounce the fifth place score out and replace it.

This should only happen if the player's score is less than the fourth place score, though, right? Keeping that in mind, place this ⑦ •If inside the •Else of another •If. We need an •[If •<•(*"Points"*) < •(*"4th Place"*)> Then], where

156 ◆ Advanced Project 4: Scrolling Shooter

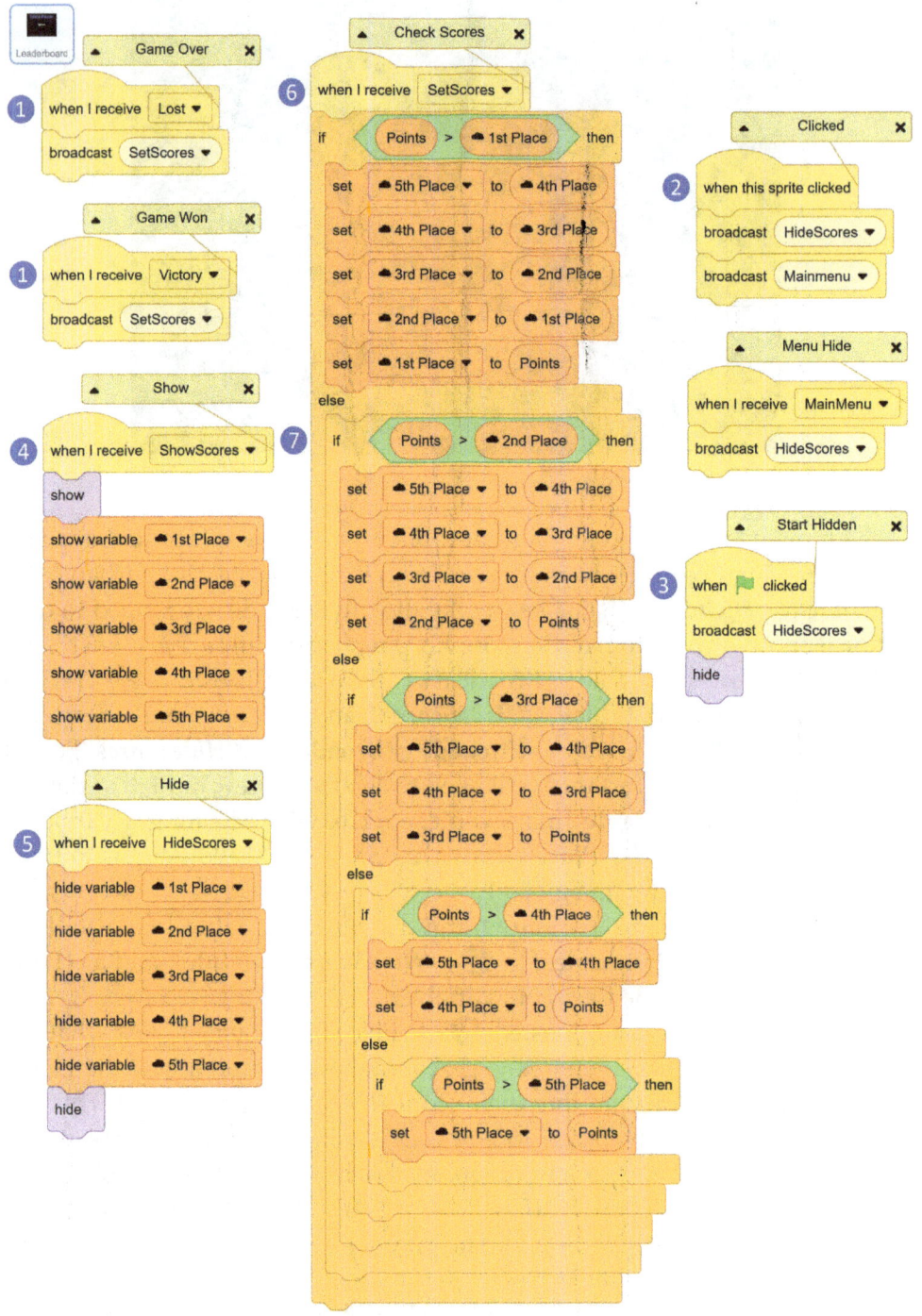

we •[Set ["5th Place"] to •("4th Place")] (so we replace the fifth-place high score with the current fourth-place high score) and then •[Set ["4th Place"] to •("Points")] (the player takes fourth place).

This happens if the player's •"Points" are higher than the current fourth place, but that should only happen if their •"Points" are also lower than the third-place score. All this should only happen in the •Else of an •[If •<• ("Points") < •("3rd Place")> Then]. Seeing how this chain of score-bumping logic works, you simply need to keep nesting in additional •If/Else blocks all the way up to •[If •<•("Points") > •("1st Place")> Then].

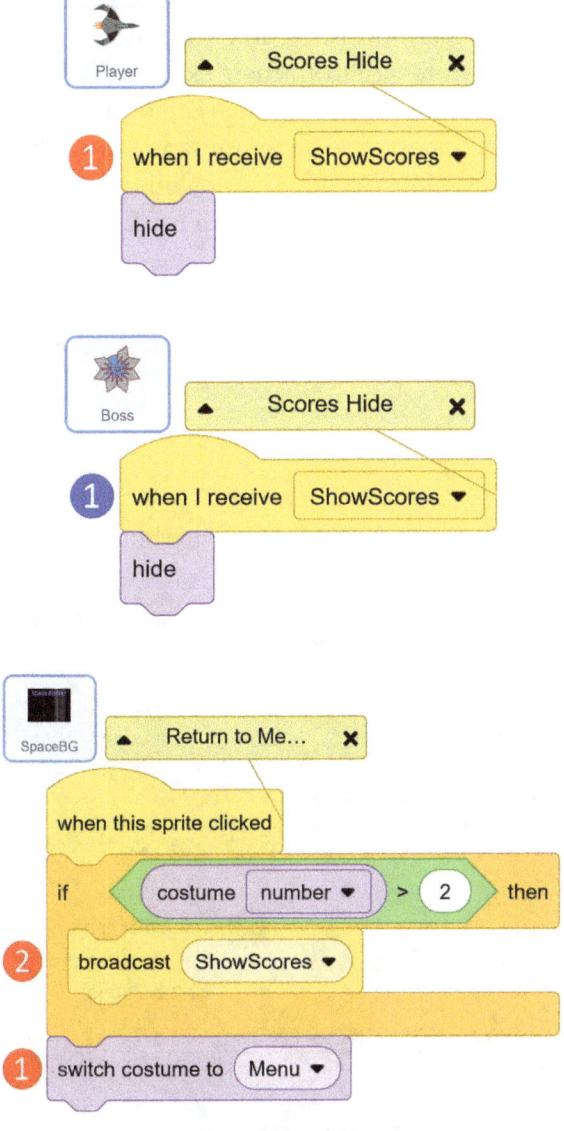

158 ◆ Advanced Project 4: Scrolling Shooter

Lastly, the ❶ *player* and ❶ *boss* need to be updated once more, ensuring that on •*"ShowScores"*, they both will •**[Hide]**. This is important because we'll jump to the *High Scores* when a player gets either a •*"Loss"* or a •*"Victory"*. To do that, go to our *SpaceBG* code on the •**[When This Sprite Clicked]** stack. Move the ❶ •**[Switch Costume to [*Menu*]]** code blocks to below the •**If**, and inside the •**If** place the •**[Broadcast]** but switch it to a ❷ •**["ShowScores"]** event. This will make it prepare to return to the menu underneath while the *Leaderboard* is displayed over it after a win or loss event.

Step 18: Difficulty Levels

Our final task for the game will be to add in a variable difficulty setting. This will allow players to experience the game in easy, medium, or hard modes to choose a challenge more suited to their skill level. This is a great way to make games more accessible. Easier settings can be a good way to let younger or less-skilled players experience the game and grow their skill, while the harder settings can keep players with higher skills interested in the challenge. We'll need a hidden •*"Difficulty"* variable to track this setting.

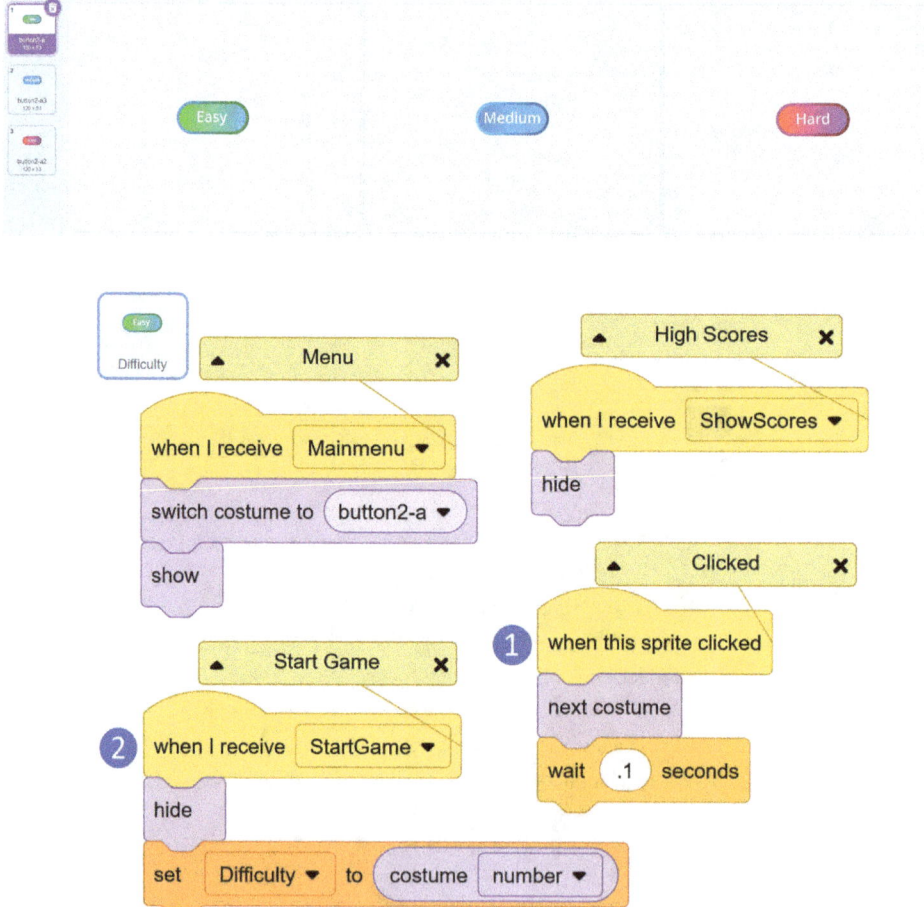

Start by adding a *difficulty* button to our main menu. Duplicate the *Start* button and rename it "difficulty". Then, in the costumes, make three costumes, each one coloured differently with one text each: **Costume 1** will say "easy", **2** "medium", and **3** "hard". In our ① •[When Sprite Clicked] code, change it to just be •[Next Costume] and •[Wait (.1) Seconds]. This prevents the button from changing too quickly. Our ② •"GameStart" code, we'll change to •[Hide] and •[Set ["Difficulty"] to •(Costume Number)].

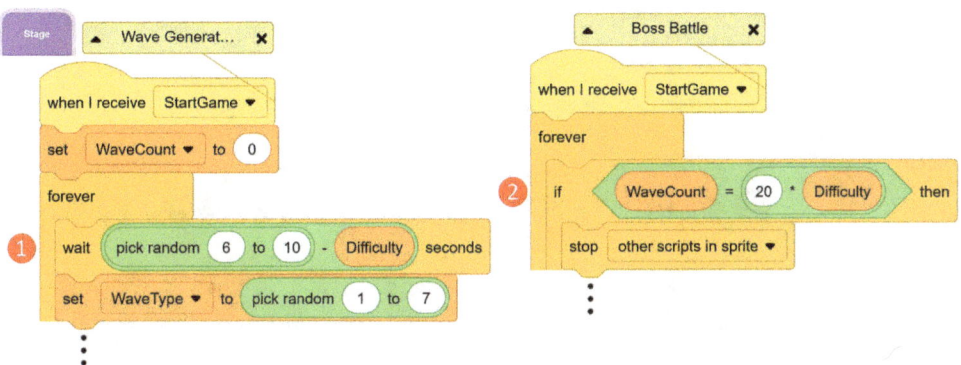

To adjust various things to our difficulty level, start with our *stage* code. Alter our wait to be dynamic to our difficulty level by switching it to a combo of ① •[Wait •(•(Pick Random (6) to (10) − •("Difficulty")) Seconds]. The harder the game, the faster the enemies come at you! It also changes when the *boss* arrives. Switch that conditional to ② •[If •<•("WaveCount") = •((20)*•("Difficulty"))> Then]. Now each level of difficulty will add another 20 waves of enemies to face!

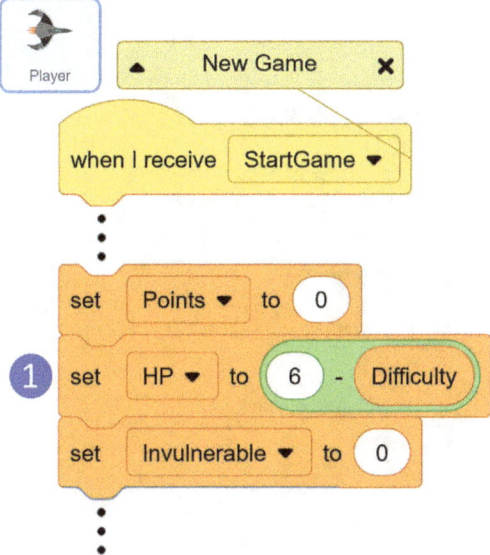

160 ◆ Advanced Project 4: Scrolling Shooter

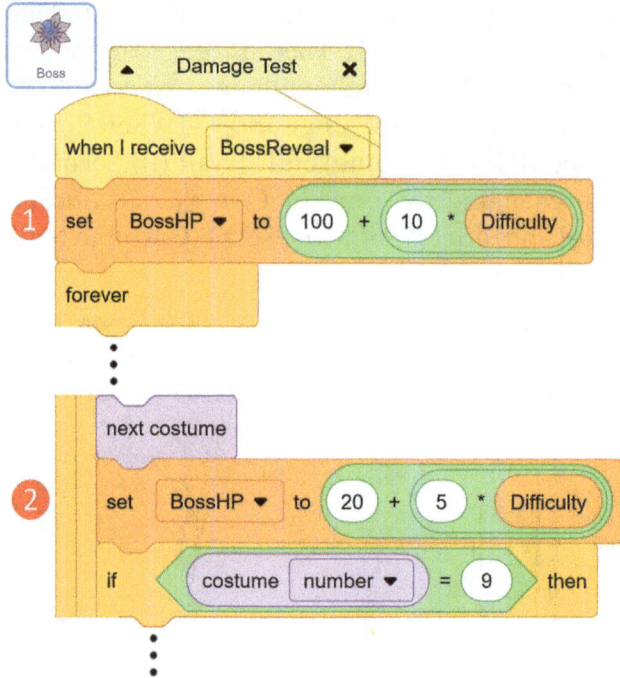

In our *player*, adjust the *player's* •"HP". Find the •[Set "HP"] and change it to ① •[Set ["HP"] to •((6)-•("Difficulty"))] so each level reduces the *player's* •"HP" by 1. In the *boss*, set their HP dynamically by changing the second •[Set ["BossHP"] to ① •[Set ["BossHP"] to •((20)+•((5)*•("Difficulty")))]. In the first one, you can make their initial state even tougher by using the same formula but using bigger numbers, like ② •[Set ["BossHP"] to •((100)+•((10)*•("Difficulty")))]. Now, the phase HP, but not the final phase HP, will gain 5HP for each difficulty level.

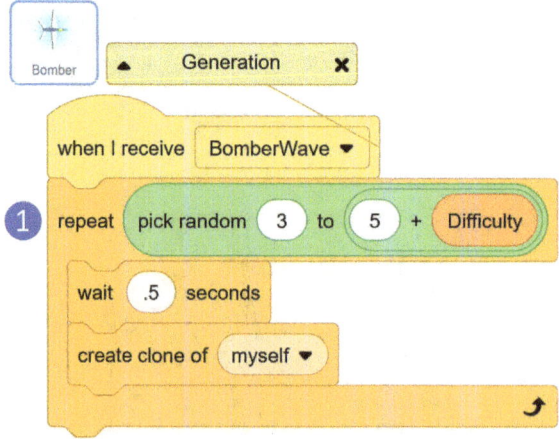

Advanced Project 4: Scrolling Shooter ◆ 161

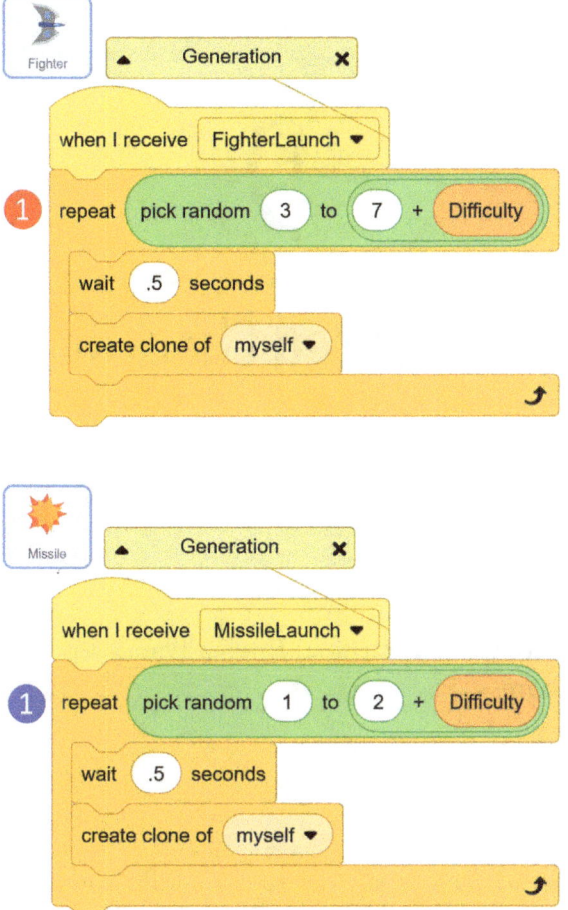

Next, alter the •[**Repeat (#)**] for the wave generation for each enemy type; waves can also be more dangerous. For the *bomber*, use ❶•[**Repeat •(Pick Random (3) to •((5)+•("Difficulty"))**]. For the *fighter*, ❶ •[**Repeat •(Pick Random (3) to •((7)+•("Difficulty"))**]. For the *missile*, ❶ [**Repeat •(Pick Random (1) to •((2)+•("Difficulty"))**]. With that, our game will adapt to the chosen difficulty level in both intensity and duration. You've now completed the final advanced project!

For working copies of this and every project in the book series, visit www.massivelearning.net for direct links to Scratch Projects, and to see our other projects and resources for coding education!

9
Advanced Check-In

Now that we've completed our final three advanced projects, let's stop and reflect on our progress.

Key Skills

By now, you should have a wide array of skills and a confident grasp of Scratch. Our final projects took our solid base of understanding from the intermediate level and expanded on it to show deeper, more fully fleshed-out projects, creating longer experiences while incorporating complex refinements and touching on more professional skills. We took our basic systems and reused or refined them to understand their potentials and showcase their multitude of uses. Hopefully, you gained an understanding of coding concepts that have you predicting design solutions, anticipating challenges, and feeling confident about your ability to build your own successful projects. In our Point-and-Click adventure, we scaled up our projects significantly and learned how to track progress, create smooth and appropriate animation cycles, and have objects influence each other to create interactive stories. We used the Platformer to learn more about level implementation, movement systems, as well as some handy progression and transition tricks. In our final Scrolling Shooter project, we learned a lot about working with clones, the basics about procedural generation, and leader board (or high score) management. We couldn't include everything we wanted in the 12 projects in this

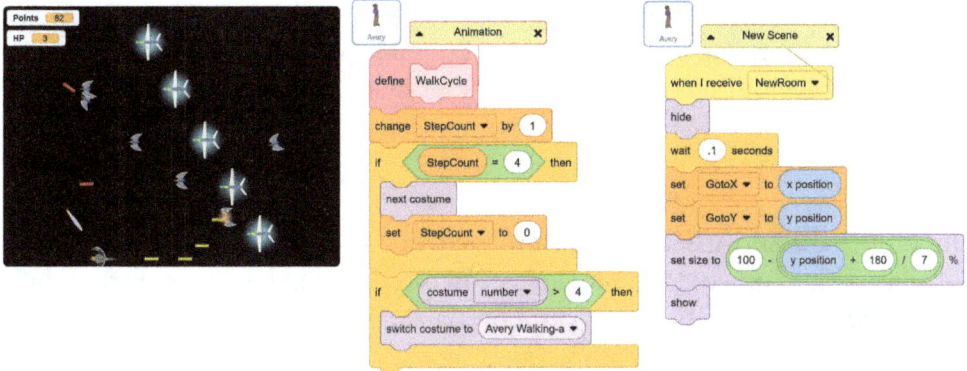

Figure 9.1 Three of the animation techniques in this book: scrolling backgrounds, walk cycles, and perspective scaling.

series but hope that you've seen enough to build the confidence to explore and try things on your own. Programming is a skill too deep to ever master; all one can do is create and continue learning.

Let's review some of the things you've learned and practised with these projects.

Animation Techniques

We learned a lot more about animation and graphics with the last four projects. Got some important practice with the art tools to make our own graphics for the Scrolling Shooter mixed with lots of coding techniques. The goal was to raise the bar on our graphics for these final four projects, in contrast with earlier games with very basic animations. These new techniques can lead to much more controlled and refined graphics.

Endless Scrolling BG

Pixel-size-specific tasks can be difficult in Scratch, but when you know where to look and know methods that work, great things happen. Because sprites rather than backgrounds are used, by duplicating this process, multi-layered backgrounds and parallax effects can be employed for even greater depth.

Stepcount Cycles

In our Point-and-Click Adventure, we implemented a "stepcount" system to time out the walk cycle animation and make sure our character wasn't changing costumes too frequently. This is a deceptively powerful tool to know and use. The same count-and-text system can run much more complicated processes and create timelines for longer animation cutscenes and other processes. For example, by skipping the stepcount=0 reset, you can keep the count going as long as needed to make timelines and set sequences of enemy generation for a scrolling shooter instead of having it procedurally generated.

Y Scaling

In our Point-and-Click Adventure, we utilized the Y position to scale objects. This is one of the many uses you might not expect for position data. By using the Y position, we can compare the % of the possible position and scale higher objects to smaller scales, forcing perspective. You can also use Y position to force layer depth, so higher objects are further back and lower objects are more forward in the layering. These two tricks can really help create a sense of perspective in a 2D game and are relatively easy to implement.

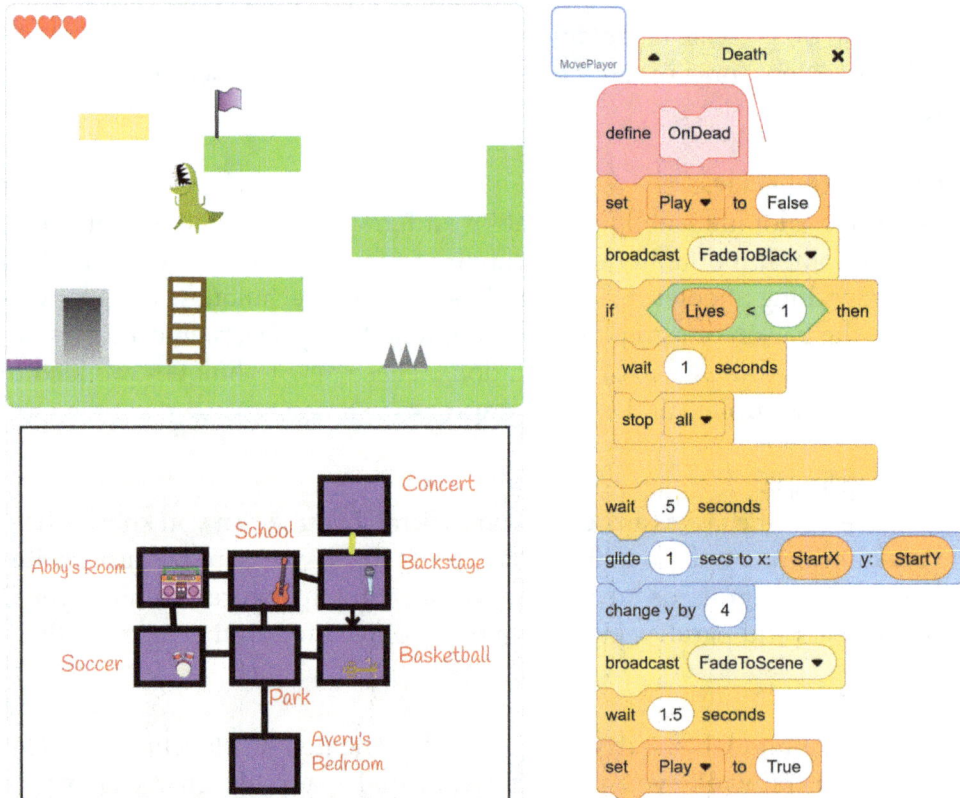

Figure 9.2 Three more animation techniques in this book's projects: movement state graphics, death animations, and pop-up displays.

Movement State Graphics

In the Platformer project, a sprite was used just to handle the graphic display for the player. Using this sprite, we tested the various situations the player could be in and set their image specific to their movement, speed, and health. For professional-level graphics, try combining this movement state system with the stepcount animation cycle technique from the Point-and-Click Adventure and you'll be working with what most 2D video games use for their character graphics.

Death Animations/Transitions

To enhance the sense of continuity in game worlds, creation and destruction should be graphically represented. Things blinking into or out of existence is a jarring experience that breaks immersion. Death animations helped with this. In our Platformer, we created a death animation sequence to reset the game when the player still had lives left. In the Scrolling Shooter, we created destruction animations for the player, enemies, and boss with frame animations and code sequences.

Pop-Up Displays

Depending on the project, you may need to display a lot of information to your user. In our Point-and-Click Adventure, two different pop-ups to display additional information were created. Our inventory system could be called up at any time, displaying overtop of the game, showing the current state of the game. A secret map was also created to help build the game and optionally be able to be kept in the final version of the game to help the player. The version on the project just had sprites on display; one could also use a pop-up's state to control whether variables are displayed.

Advanced Variable Handling

We learned to use variables in a number of useful ways to create new control systems, work with clones, and save data. Computers are fundamentally data-processing machines, and variables are data. Tricks and techniques for dealing with variables are a huge part of computer science and coding, and while it's true that there's so much more to learn, we've got a great tool set after working on all these projects.

Limit Systems (Ammo, Timer)

In all projects, conditionals were used to limit activities based on variables. In the Point-and-Click Adventure, key/gate systems took care of inventory items. In the Platformer, having player lives demanded more cautious

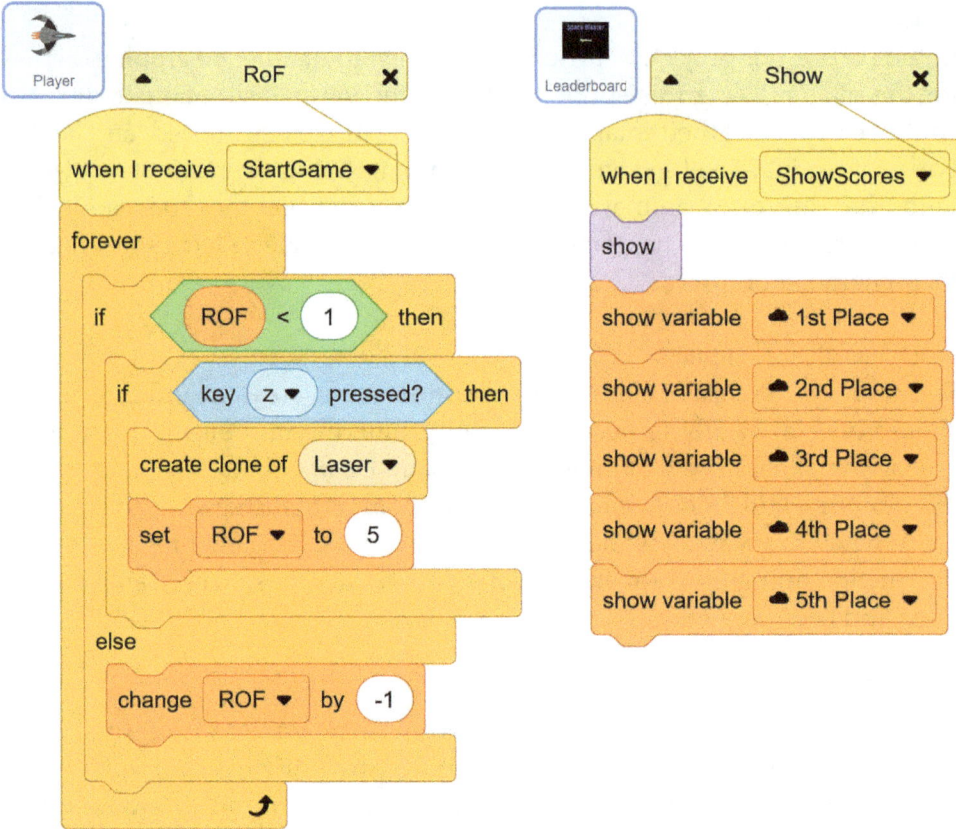

Figure 9.3 Two useful variable techniques for aspiring game developers: a ROF, or rate of fire limitation, system on the left. On the right a leader board sprite making the top five scores visible.

gameplay. In the Scrolling Shooter, timer limits were used to control invulnerability, the rate of fire, and when the boss would appear. Also, there was a random chance limiter on the rate that enemies would shoot and what enemies would appear.

Child, Parent, or External Properties

We worked a lot with clones in our Platformer and Scrolling Shooter and had to deal with the need for pushing data to them. Universal variables were used to determine moving platform movement to create external control, allowing one variable to control the behaviour of multiple objects. In the Scrolling Shooter, costumes of clones had their own unique individual hit point count, where a variable wouldn't have allowed individual tracking for each clone. By default, all clones have the properties of their original sprite, or parent.

List Variables

In our Bar Charts and Data Files project, we learned to do some basic data sorting and data visualization using a list variable. This advanced type of variable is also known in computer science as a data structure, basically a bunch of variables that relate to each other in some way. We learned to step through the list in order so we could assess or process the data in order. In Book 2: Intermediate, we had a preview of this by learning to create a map grid, but we didn't have a proper data structure to work with. With proper data structures, we can use more code blocks to work with complex data more easily.

Cloud Variables

For our Scrolling Shooter, Scratch's cloud variables were used for the leader board. This allows the game to track high scores entered from any player that plays it. Cloud variables are a fun way to let you let players influence each other. While we used it for the fairly obvious use of a high score leader board, you can put these to other clever uses wherever you might want to capture player data. For example, you could have a racing game keeping track of how many kilometres all players raced collectively, or how many games have been played, or won, globally.

Leader Boards

Leader boards could be created with local variables or cloud variables. It can get confusing handling high score tracking with a lot of nested statements, so we created a sprite to handle it all for us. With multiple difficulty ratings in the game, one could worry about the fairness of the high scores. Since the game gets longer based on difficulty, it will always generate much more potential points for players.

Complex Movement

By adding in more conditional tests, influences, adaptations, and iterative functions, we ended up with much more sophisticated movement systems. We learned to adapt and interact with the worlds we created by enabling more dynamic and reactive movement and characters.

Jumping Movement

For our Platformer project, by using ground collision tests, we made sure our player had to be standing on the ground to jump. Our speed-based movement meant we could use variables to create factorial effects and arcing jumps.

168 ◆ Advanced Check-In

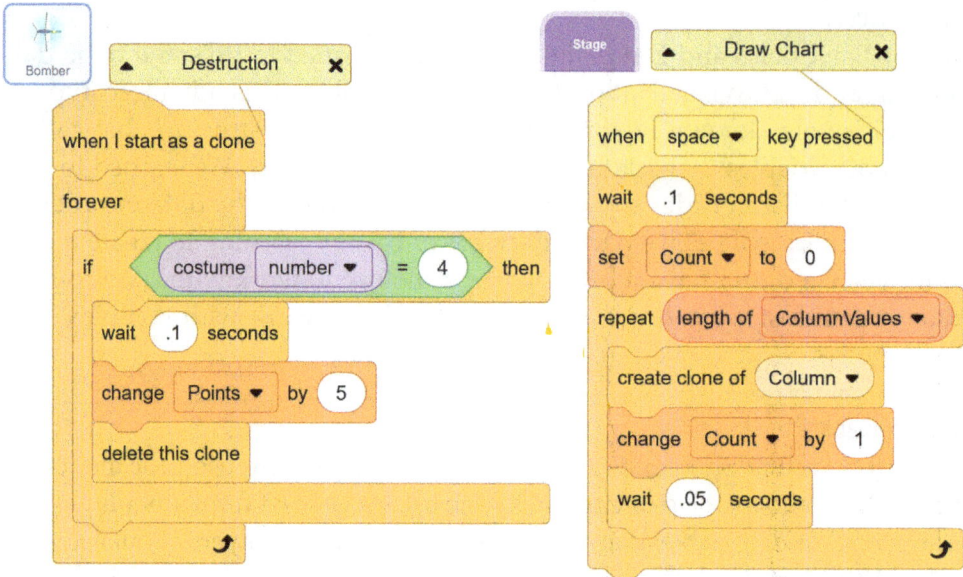

Figure 9.4 Two more useful variable techniques. On the left, using costumes for HP from Project 4: Scrolling Shooter. On the right, an example of a "list walking" algorithm that uses the length of a list variable to work through each entry in the list.

Inertial Movement

With speed variables, movement based on speed can be created, and with iterative factoring of speeds, a system for inertia and friction came to life. This was used for gravity and running. It could also be incorporated into terrain systems to allow for sticky or slippery surfaces.

Terrain Interactions

In our Platformer project, we created a number of interesting types of terrain for the player to interact with while being mindful of important and hard-to-realize strategies for dealing with terrain, especially colliding with it. Side-view projects like our Platformer make it a bit more difficult to determine terrain collisions, but we've seen multiple cases of working with it. To test out the concepts, one could try incorporating some terrain into a top-down view game like our Big Map Racing project from Book 2: Intermediate.

Guide Objects

This hard-to-imagine but powerful technique was used in the Scrolling Shooter to have enemy units be able to both spawn and move to dynamically calculated positions. This technique can make some very complex behaviours possible with minimal effort if one gets used to the concept.

Inherited Movement
One of the more complicated things we did was allow objects to influence each other's behaviour in the form of moving platforms. The player would detect any moving platform they were standing on and inherit its movement in addition to their own. Inheriting movement or other factors can be a very convenient and powerful technique for creating reactive and interactive worlds with moving platforms, conveyor belts, wheels, gears, and other machinery concepts.

Movement Patterns
To create our final boss in the Scrolling Shooter, a basic movement pattern system was used. Here, the boss would randomly select a tactic, and then, through a series of conditional tests, it would acquire the appropriate movement programming. This same setting/pattern execution system can be used for all kinds of changing behaviours to create richly reactive and interactive characters and machines. It can be used for lots of behaviours besides just movement.

Collision Handling
Collisions, or being able to know when two objects occupy the same space, are a huge part of building interactive models and games. We learned a number of new ways to test and use collisions in the last three projects. In our Point-and-Click Adventure, collisions were used to control scene transitions with our exit objects, but also for handling our pop-up objects, having them receive click collisions instead of the game underneath, creating a user interface separate from the game. In our Platformer, collisions were employed extensively to test how the player should be moving with gravity and terrain effects determined by collisions with the ground and other objects. In our Scrolling Shooter, our combat was ruled by collisions testing for weapon hits, vehicle collisions, and explosions effects.

Collision Masks
To better handle the play dynamics of a platformer, an advanced technique called "collision masks" was used. The technique involved using one object to detect collisions to determine movement, and another to handle the actual character graphics. By not using the display graphics to determine collisions,

Figure 9.5 Three movement examples from Project 3: Platformer. At the top, the jump action script with its important ground-testing technique. In the middle, the method for inertial or slippy movement, and at the bottom, the collision testing method.

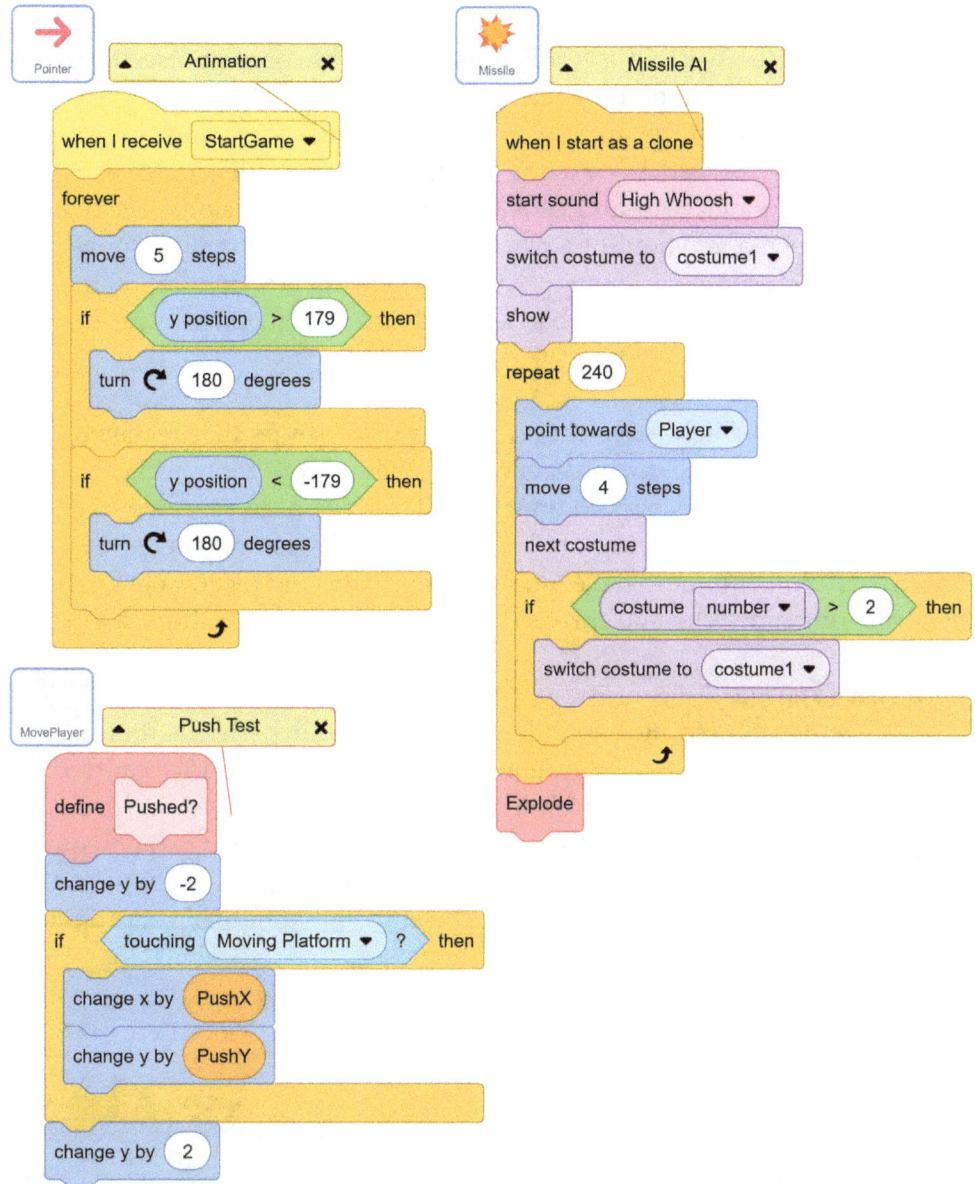

Figure 9.6 Three more examples of movement techniques. At the top, the pointer object that is used as a reference for many objects in Project 4: Scrolling Shooter. On the Bottom, the push test for the MovePlayer sprite in Project 3: Platformer that allows inherited motion from a moving platform. At the right, the missile sprite from Project 4: Scrolling Shooter and its heat-seeking movement pattern.

but rather another object and its costumes, we could control exactly how the player interacted with the world while still having total freedom with the art. Collision masks are a very powerful way to control interactions while freeing your art from the risk of interfering with interactions. A great, more complex example can be seen in fighting and action games' hit box systems.

Collision Testing Movement

In our Platformer, we had a lot of constraints and consequences to movement we wanted to simulate. There was the need to making sure the player wouldn't fall through the ground, could stand on ladders and move up or down them, get moved by moving platforms, and could stand on or fall through blinking platforms as needed. We did a lot of collision tests to determine what the player should be reacting to and controlling their movement through it. We used collisions to "unmove" players that collided with solid objects in order to have reliable barriers and other interactions as needed.

Clicking and Collisions

In our Point-and-Click Adventure, we created pop-ups that received click events, and when done, the objects behind them didn't receive click events. Similarly, when players clicked on sprites, the stage didn't receive a concurrent click event. The layering of objects controlled what registered the click and what didn't. We could have the stage or background receive clicks as well as sprites, but only when no sprites collided with the mouse click on top of it.

Progression Systems

Unless we're just building a singular self-contained experience, there's need for some kind of progression system to allow the user to move on to other scenes or experiences in our projects. We need not only a system for saving the progression but also a way to handle the transition or change that required to enact that progression. There's a lot of different methods to both present these changes and track those changes presented in our advanced projects.

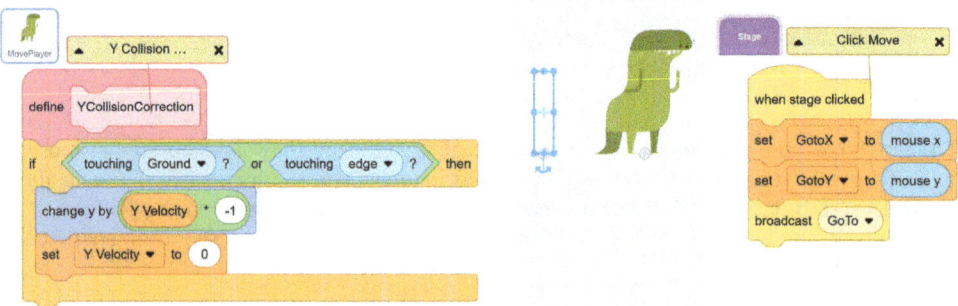

Figure 9.7 Three examples of collision techniques used in this book. The Y Collision technique from Project 3: Platformer's MovePlayer sprite. A comparison of the two costumes of the MovePlayer and DrawPlayer sprites from the same project used for the collision mask technique. And lastly, the movement target–setting technique when clicking the stage in Project 2: Point-and-Click.

Player Lives and Restarts
One of the fundamental techniques of video games. We used a set number of lives, with a number of dangers potentially costing a life if unavoided in the Platformer project. The player could retry the same level if they had lives left but would lose the game and have to restart if they didn't.

Levels/Scenes
Dividing gameplay into distinct scenes or levels is an almost-universal aspect of games since all but the earliest days. The Point-and-Click Adventure project had distinct scenes the player would move between. We learned how to make a transition system to control movement between scenes. In our Platformer project, distinct level designs were created, then a door sprite was used to allow players to move to the next level. We learned a few ways to transition the game between levels, ensuring all objects were properly aligned to the correct level state – we used costume changes for the ground, single object repositioning for the ladder, and clone spawning for the moving platforms. Providing you with working examples of three different systems.

Two-Way/One-Way Transitions
In our Point-and-Click Adventure project, "exits" were used to change scenes. Most exits were paired in the next scene to allow the player to move freely back and forth between adjacent scenes, but one was a one-way transition as an example. In the Platformer project, transitions were handled by entry and exit doors but were only one-way, so players couldn't return to earlier levels, only press forward.

Level Cloning vs. Level Costumes
In our Platformer project, we used a number of different techniques to handle level transitions. For some sprites, we did level-specific costumes, a very simple and straightforward technique. But in some cases, an object might need to be able to have animations or use its costumes for other things, or each instance of the object in the level may need to interact separately. For those cases, a level-specific clone system works better.

Waypoints
Waypoints allowed players to reach a specific location and have their progress saved. If they lost a life, they'd normally have to restart the level from the beginning, but if they reached the waypoint, they'd restart from that advanced point instead. In the Platformer project, the waypoint was also used to unlock the exit, adding a level of secrecy to the level design as well as a progression impact of the sprite.

Figure 9.8 Four examples of progression systems: the player lives display, waypoint collision, and level-specific costumes from Project 3: Platformer, as well as an example of the inventory system used in Project 2: Point and Click.

Inventory States

In our Point-and-Click Adventure, we had a basic inventory system consisting of a number of individual variables. Each variable remembered (kept track of) the state of an object in the game story. As the story progressed, these variables would be changed by testing conditional statements to match the story progression. We didn't use it in these examples, but these inventory variables could be coalesced into a single list variable, but with our Bar Charts and Data Files project, we showed you some basics for working with lists, so maybe you can try that in a project of your own!

Key/Gate Systems

In a "key/gate" or "key/lock" system, an object or variable is an earned or acquired prize that unlocks progression. Be it some area, item, or other gameplay dynamic, it is locked away until the player gets the key to open its "gate". The player wins the Point-and-Click Adventure by getting all the instruments, having the final concert "gated" until all these "keys" are acquired. In the Platformer, the exit is "gated" until the player gets to the waypoint, which acts as a key to unlock the exit. This system can use variables, individual values within variables, list variables, or through events, even no variables at all.

Play Dynamics

A lot of what we explored in these final four projects were about game or play dynamics. While our focus might be education, we shouldn't underestimate how useful building games can be to educating students. These systems help make interactive experiences that will engage people, while also being powerful systems that students want to experience. By tapping into their interests, they'll value the work and have more drive to do it. They'll have self-interest rooted in self-expression to power them through the hardships and inevitable setbacks in project development and coding, and everyday life.

Difficulty Scaling

A difficulty system allows various aspects of the game to be altered, making the game easier or harder for the player. For our last project, we just used mathematical formulae to incorporate a difficulty-based modifier. We could have used difficulty to change the setup or allow or disallow actions or components in the game. This could add or remove enemy or power-up types or allow enemies to do new things, include additional levels, or just about anything.

Procedural Generation Basics

For our Scrolling Shooter, we used a very rudimentary procedural generation system. Spawning enemies on a randomized timer, of a randomized type, at randomized positions, and in some cases with randomized movement goals. This is about as simple as procedural generation comes, but it's a great start for students. You could challenge advanced students to create formations of enemies instead of the steady stream creation that just makes left/right sweeping lines. The key is that students will see how we can use computer systems to generate content rather than having to create explicit placements and timings for things manually, and they can hopefully explore both concepts.

Clone Waves

By using the original sprite as a generator object, we could move it as needed and create clones with them inheriting position and other properties. By controlling the creator of clones, new patterns, clusters, waves, or phases of clones can be created.

Dynamic Spawning

We took our cloning to another level by having a clone cloning objects. This can be very dangerous, with Scratch crashing if you try to create too many of them. You need to be very cautious of having clones you didn't think about creating, or iterative chains of cloning if clones are creating clones of themselves. In our case, our clones were creating lasers, but we had the challenge of needing them to position and orient themselves, but they would inherit only the original object's properties, not the clone that creates them, so we had to learn to use variables to store values used for creation.

More Advanced Practice

With all 12 of our projects completed, you should be feeling pretty confident and comfortable with Scratch. However, there's always more to learn and try in programming. Even after 30 years of programming, I find that I have yet to learn tons and many ideas I haven't tried yet. You should always look to expand your knowledge and skills. At a basic level, teaching may be about properly teaching techniques, but truly great teaching inspires students to learn and seek out their own path for the rest of their lives. Ideally, this book inspired you to continue to learn programming, seek out new ideas and

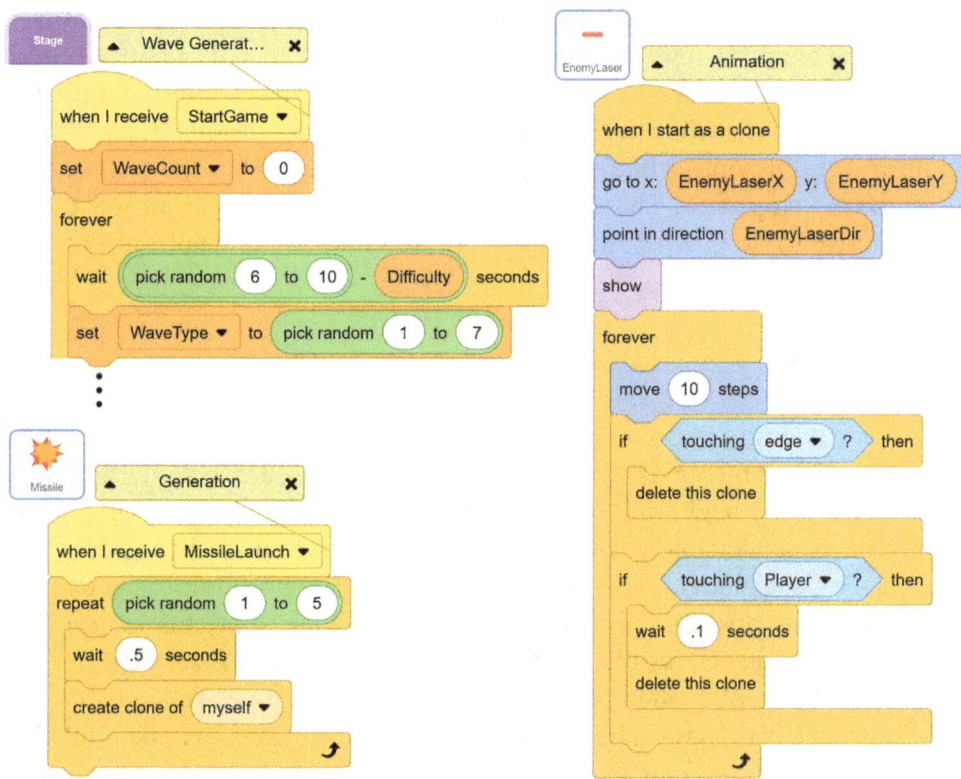

Figure 9.9 Three important code examples from Project 4: Scrolling Shooter for creating various play dynamics. On the left side we see some of the stage's enemy wave generation script and the missile sprite's reaction to a launch. On the right, the *EnemyLaser* showed how to get clones to generate their own projectiles accurately using stand-in variables.

concepts, and challenge yourself while continuously creating. The following suggestions are projects you might want to try creating. They are common projects on Scratch, usually very popular with students, and give a great opportunity for continued learning.

Pet Sim

A pet sim, or pet simulator, is a project that creates one or more virtual animals that the player can interact with. It generally includes some aspects of animal care, such as feeding, cleaning, exercise, and some degree of interaction/relation-building. Many include the ability to get things to improve their environments, decorate, or expand experiences. Some go even further, expanding into fantasy territory with the pets being mythical creatures, with some way of using the creature in adventures, such as a game experience or competition like Pokémon. These projects can start very simple but often greatly expand with the use of store systems, various game modes including minigames, data saving for both the design of the pet, and its growth/development. These projects work great not only to practice coding but also to tie in with storytelling, or any number of health topics.

Arcade Classics

To anyone wanting to pursue a career in game development, I always tell the same thing: start by remaking a classic Atari or arcade game – something very simple, from the beginning of the video game era. These games are simple enough to be achievable but still provide some important challenges. By having a very clear model (a completed game), one can work to pursue as close a recreation as possible, preventing one from taking shortcuts or ignoring important aspects. Games like Space Invaders, Breakout, Asteroids, or Centipede can be great challenges to recreate and offer a lot of fun as a reward. While using Scratch, we'll have some definite adaptations to make, but classic games can be a great way to challenge advanced students or yourself, and they can be great bases to work from to create new adaptations of well-established genres. These are great for practising clone creation and management, collision handling, and data tracking for scores or levels.

Life Simulation

The Game of Life is a computer science classic, an investigation of complex emergent systems using cellular automata. The basic concept is to make small "cellular organisms" that have very basic and rigid behaviours, but through having them propagate, live, and die, we can see how they interact and interfere with each other to create complex ecosystems or machines. A DaisyWorld simulation is a similar concept, generally with global environmental factors determining growth rates of black or white daisies, which in turn influence the global environmental factors. Simulation systems like these are a great way to explore computer science simultaneously, acting to illustrate and bring to life mathematical and science concepts that rule our universe. By exploring them, we can practice our data manipulation and collision systems but also see feedback loops and algorithms spring to life on our screen and help create deeper understanding of complex systems and the concept of emergent systems. These are wonderful tools to deepen our skill set and spark wonder and learning in our students.

Teaching Advanced Scratch

Adding coding to the curriculum is perhaps the most daunting challenge ever put upon teachers. I think most teachers will come to the subject with a lot of trepidation, if not outright fear, due to its complexity and foreignness. Hopefully, this book series and the range and scope of its projects provided you a sense of familiarity with Scratch and coding concepts that has helped see the value and potential in coding. The advanced projects in particular should be appropriate for grades 8 and up and give a great footing to move on to text coding with a couple years of practice at this level. By mastering the skill to this level, you should be able to comfortably analyze coding projects, prepare lessons, see cross-curricular examples, and continue to self-develop. In an ever-digitizing society, these skills may be boundary-pushing today, but in a decade, they will be a fundament of education.

Teaching Comfort

Perhaps the most important struggle we face in this revolution in education is getting teachers comfortable with teaching coding. We're asking an entire profession to jump on board with teaching a completely foreign subject that has the technical rigours of mathematics and the indefinite answers and need for value judgments of English. This is an incredibly challenging request that society has placed on educators, and it's absolutely understandable that this change would be challenging at best. With no added time in the day, rarely any chances to meaningfully study or train in the subject, and often a tremendous lack of resources to support this transition, it can be a nightmare to approach. Coming from the industry and seeing both the need for employees and the power of technology to transform lives and society, I want to help solve this impasse of educators moving into this new territory.

Our earlier projects helped introduce the code blocks; by our advanced projects, there wouldn't be too many blocks that wouldn't be obvious, if not familiar. We touched every aspect of the Scratch editor, so navigating around it should be second nature by now. We worked on a few projects where we needed to make the art ourselves and should have relative comfort with the Costumes tab's art tools. With this foundation established, we then started seeing how code blocks interact with each other and started understanding the combos – how to build If statements, how to build math formulae, how to work with loops. Through our beginner and intermediate projects, we established this elementary knowledge, then in our intermediate and advanced projects, we really started learning the techniques and theory.

Our goals aren't just to be good coders ourselves, though, so knowing isn't enough. Ideally, you'll get familiar with Scratch to the point to predict student's possible mistakes. It's hard to get this level of knowledge without first teaching it. You'll find all kinds of errors when you start teaching coding. Teach enough and kids will introduce you to just about every error possible, if you haven't found them all yourself first. It's important to know that errors are a part of the process. *"If you aren't making mistakes, you aren't trying hard enough"* is an old saying that definitely holds true with coding. Humans simply don't think like computers, and never will. Our goal isn't to think like a computer but to predict some of our common mistakes or to be able to interpret errors that happen. We can step back and look at things from a computer's perspective to analyze our code when needed, but our human creativity is a precious thing that computers simply can't replicate. And that's something to be celebrated. Our troubleshooting chapter should help you recognize and deal with the most common problems so you can stay focused on the creative side, not be bogged down by errors, and don't have to worry too much about when things go wrong. Rather, embrace them as opportunities for learning.

Planning and Designing

By getting more comfortable with the techniques and theories, one can better imagine, plan, and design projects and lesson plans. You may want to explore the design thinking model – empathize, define, ideate, prototype, test, implement – with your students to focus on the planning and design process, but it isn't necessary. For younger grades, we recommend more direct-led, pre-developed projects in order to build their familiarity with the tools and techniques before having them design too much. Template projects can be a great way to give them a framework to fill with content to their own interest and goals. This can give them some structure to work from and take some options off the table to help them focus and coordinate. Intermediate and advanced students can increase their design freedoms, take on more challenges, and use less guides and templates. This helps them transition from being led to being leaders.

As we get to advanced skills, our students should be able to plan and design projects dealing with a number of factors: object, variable (or data), state, flow, limit, and trigger strategies. We can get them to articulate their ideas with definitions, statements, and plans about their goals and their projects', and components', purposes. We can have them plan and describe what objects, variables, and states they'll need and how they will work. They can recognize, define, and plan for game states, process flows, limiters, and other controls, as well as event triggers. Not every project will go according to plan, and that's okay; they don't have to be completely accurate in their plans and predictions, but we want to highlight how they are going through the thought processes and, importantly, can both verbally and in writing share, explain, and defend their ideas. Communication is an absolutely critical part of tech, and we should ensure using our time working with coding to reinforce other critical skills. We can allow students to improvise and adapt as needed, but by having them plan, we can refine their understanding and help prepare them for a world that demands communication, documentation, and teamwork.

Testing and Reviewing

Perhaps the biggest challenge to teaching coding is being able to test, review, and grade our students' work. Without a firm grasp of coding, how can one assess code? Hopefully, the training this book provides has helped on that front, but there are plenty of other approaches to help beyond the obvious need to increase our subject expertise. Our advanced projects showcased some important things to look for or consider in testing projects. Problem areas and consequences were highlighted in a number of ways that you can look for in student projects. We can also analyze our own projects and lesson plans for these issues and try to prevent coding them when putting together our own teaching materials.

We consistently waited to place any kind of menu in our projects until very late in the process. This meant we often had to go back to objects and change some things to accommodate the new menu. We did that to shortcut the process. Menus mean delays when testing. Shortcuts are key to making development a smooth and enjoyable process. Can we skip ahead and get to the core of the project to make sure the fundamental systems are working? If we need to test a specific level, can we simply skip ahead? These development practices and tools can be very handy for making the whole process efficient. Whether they're included in the final build or not is a different matter. But if projects are built with convenient shortcuts, it can save a lot of time. You can have students leave them in for you to use and test as well, or as a compromise, keep the code blocks and disconnect their event so you have to reconnect them in the editor to enable them.

At younger grades, we can simply play projects and see if they do what was intended, or even run at all. As students increase in skill and projects increase in sophistication, we have to start being able to be more discerning in our analysis. As we showed in the Point-and-Click Adventure, how conditionals are chained together, either nested or sequential, can have a big effect on how they run. When we see our students using conditionals, we want to double-check what all the logical operations would be. Did they account for everything? Are they getting unintended results due to incorrect nesting? Did they forget to account for some potentials? Similarly, we can see if they account for any values in their project. If they allow a value to be set to given ranges, does the program respond appropriately to all those various values? Can values slip outside of expected limits?

Also look for other exceptions in programs. Do objects activate, deactivate, show, and hide appropriately? Do objects layer correctly? Do processes run in sequence, or concurrently as appropriate? Do things trigger correctly consistently? Scale your critiques to the level of the class and the students' skills, looking more critically as the need arises to keep pushing them forward in their skill development. The goal is not to be pedantic but rather to push students forward. If you're suspecting reuse of code from other projects on Scratch, you can look for how well-adapted that code is to the unique use within their project. If they didn't work to create the code, did they work to adapt it and make it appropriate to their needs?

As well, don't be afraid to recruit your students into the testing and reviewing process. Quality control or testing is a major area in tech; no product or service can survive without it. Get your students to do peer review. Challenge them to find bugs or flaws in each other's work. The key is to foster a positive culture around this. Reviews and critiques are critical to becoming the best we can be. This process helps us become better creators, not just by getting feedback, but also by having the opportunity to review others' materials too, to see things from both perspectives. This can help you and the student get to hear a diversity of views and viewpoints on a student's project as well, especially important since they probably didn't think to make a project aimed at a teacher audience!

Analysis and Explanation

Coding is an interesting collision of logic and language. We call them programming languages because they're how we communicate with computers. Coding is about learning to communicate with computers so we're able to tell them what we want them to do. An essential skill is being able to read this language. We, and our students, need to be comfortable with reading code as it is a key part of doing analysis. This practice helps us work through the sequence and purpose of code. This can be a key bug-hunting skill for one's own use but also allows us to try to understand each other's code. As teachers, we absolutely need this skill to evaluate our students and guide them to finding errors.

Reading code starts with simply being able to compare students' projects to the exact code of a lesson project but eventually expands to understanding the meaning of code and to being able to predict the outcome of code we read. We want to become fluent enough at code to predict its results, but that is a somewhat-impossible task. Eventually, as the level of complexity in projects grow, it makes them too difficult to understand all at once. When that happens, we simply have to follow one object or script at a time. As we expand our ability to comprehend this language, we can best predict behaviour, but in reading code we become more familiar with the processes, methods, and techniques. We increase our ability to learn from others by reading their code. Reading others' code is both one of the hardest skills and one of the most valuable to have.

But analysis doesn't simply come from reading code. We can also learn to interpret the result of code from the output. By simply using programs or watching them, even without the code visible, we can start to imagine the code behind them. What events are being triggered? What are the unique objects in play? How do they interact? What conditional statements are being used? We begin to make sense of experiences through the logic of code. We predict how they could be built or at least how we would try to build them. We can explain the phenomenon we see through logic and object-property values. This allows us to understand the world through computer science and information theory but can also help us learn to communicate about coding and projects to help guide students make their projects better.

10

Follow-Up: Extending the Projects

You may have noticed while building the projects that some things could be done differently. Maybe a feature you thought should be there was missing, maybe the myriad of techniques we showed could be simplified to your one favourite method, or some other changes. These projects were built with plenty of room for improvement in mind. Some were simplified to fit the learning curve, or to provide a diversity of concepts, or even to fit a more linear path of development while retaining usability at every stage. There were a lot of considerations to make while building the example projects, and having them calling out for you to make them your own was always planned for and intentionally put there to inspire you to try improving them.

A lot of great development is iterative. We build a system, get it working, then go back and improve upon it. That's what this chapter is about, giving you ideas to go back through all the projects and add new functions and features to rediscover them with the knowledge you gained since doing them the first time. This can be one of the challenges in coding. As we learn more or even just change tastes and preferences, we can look at a project and want to use totally different methods to achieve it. We'll run through each project and give you some examples of ideas and features that you could explore adding to the projects, but only in loose terms; it'll be up to you to figure out how to pull them off and make it your own. Even if you don't know the exact techniques for some of them, you should have the familiarity and expertise to take on the challenge now. It's time to strike out on your own and conquer your own coding mountains.

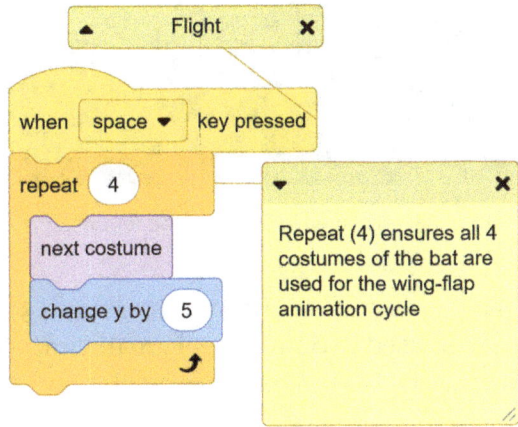

Figure 10.1 Comments can be collapsed (flight) or expanded. When expanded, they can be resized to any shape by dragging the bottom right corner. Switch their state by clicking on the arrow in the top left corner, or delete them by clicking the top right "x".

Commenting

When we go back to our old projects, we can find them harder to understand than we imagined because we've changed our own thoughts and perceptions as we've grown. If we remember to think of coding as language, we can perhaps understand the change in more human terms; who as an adult still speaks like they did when they were a teenager? We learn new phrases, get new catchphrases, settle into new habits, accents, or mannerisms.

Because coding requires us to both translate and interpret as well as keep a mental model of the project or processes, it can be very difficult to read other peoples' code, and our own code if it's been too long since we last laid eyes on it. One habit we should get into, along with our students, is commenting. Commenting is the practice of providing text hints and descriptions added to a project to explain things using natural human language. Comments in Scratch take the form of sticky notes, either at large or attached to specific code blocks. Simply right-click to add a comment on the background for an at-large comment, or on a code block for an attached comment. You can resize comment notes as needed by clicking and dragging their bottom right corner. You can see I used comments to label all the stacks in my projects, but this is just one minor use. Good practice would be to be more descriptive and especially mention any connections with other code or objects.

Try to go back through the projects and add comments to describe what stacks do, what variables are for, what objects will react to what broadcasts, etc. Add notes to your projects that let you read and easily understand its structure, flow, and processes at a glance. This is a key skill for professional

Figure 10.2 The four beginner projects covered in Book 1 of the series.

programmers and helps immensely when two or more people are handling the same project. Getting used to doing it will greatly improve students' skills while making evaluating their work easier as well.

Dinosaur Dance Party

We've come a long way from our first project. The original interactive musical animation was a good introduction, but there's so much more we know how to do now. We already suggested a lot of tweaks to try in Book 1 and 2, so here are some really ambitious takes on redesigning a music and animation-themed project.

If you add some lyric displays to the project to follow a song, can you add in a karaoke-style timing indicator? You'll need to figure out timelines and animation spacing and timing. With a timeline system, you could also make a special effects and animation show that matches along to a whole song, not just a repeating loop. If you let the user switch songs, can you switch up the performers, stage dressing, and their behaviours for different songs? Can you add in some other controls and behaviours to let the user interact and express themselves to the music?

Fireworks Display

Revisiting our Fireworks project, it clearly cries out for using clones to generate fireworks so they can be reused. If you used a custom My Block, you could even use parameters to modify how fireworks are launched. Could you set up a launch sequence to call out the different fireworks in a large show without requiring more than the initial launch instruction from the player? How about creating a compound firework – one that launches and, when it explodes, launches out other smaller fireworks? You could celebrate your national holiday with some pop-ups during the firework show to talk about the history or feats of the nation during the show. Since we never used the date/time sensing blocks, maybe you could try to use them to calculate how long until the next national holiday, or until New Year's Eve.

Batty Flaps

Again, a game with a very simple premise can be hardest to adapt, but you can look to other games in the genre, or similar genres for inspiration. If we look to other endless runner games, we could try obstacles that require different reactions to pass with different character states. Some running-based games have obstacles to jump over, or slide under, needing timed actions, not just positioning. Perhaps the bat could fly away into the background or forward toward the player to avoid some obstacles? Or maybe they have to use a dive or other manoeuvre or attack to break through certain objects or barriers. You could even make different creatures to play as, who maybe even behave differently. Of course, clones would be the ideal way to create the gates, and you could implement a scrolling BG. You could also look to old 2D submarine games like Sea Dragon (I recommend the ZX Spectrum version) for a totally different take that helped define the concept of both a "scrolling shooter" and an "endless runner" game.

Butterfly Catcher

Our first game project was pretty simple, but effective. Sometimes simple games are the hardest to re-imagine or expand on. In this case, we have gameplay changes we can make, but we also didn't do any of the polish on this game. So of course one could add in a menu and Game Over, but you could also add in global scorekeeping with cloud variables and a leader board display. You could add in difficulty levels that alter the number of lives or the relative sizes of the sprites. While the game automatically scales for difficulty, you could add to a menu system to allow choosing different themes that switch the background and object costumes. You could add in a timing system, so maybe you score more the faster you are, or you automatically lose a

Figure 10.3 The four intermediate projects we covered in Book 2.

life if you take too much time. You could add power-ups to change objects' sizes or speeds temporarily. You could also add in some "juice" with some special effects like sparkle bursts when you catch the butterfly.

Pen Tool Fun

In the original project, we only used a single object creating our drawing, but any object can use the Pen! What happens when you start having multiple objects drawing simultaneously? Can you make some fractal drawings by using clones that use Pen tool code blocks? Even just sticking to very mathematical tasks, there's still tons we can add in and explore. Can you create an animation with the Pen tool – having it draw something that is changing over time? What about having the variables hidden and only show when you move the mouse to an edge or corner of the screen so they don't obstruct the view?

While our Pen Tool Fun project was pure drawing, don't forget that you can use the Pen tool in other projects; it can be an interesting way to do special effects or even be a part of gameplay itself! Any colour you draw can become part of the terrain of a platformer, or you could programmatically draw out a maze. You can revisit this project by using what you learned in it and applying it to your other projects in interesting ways. The **<Touching Colour (Colour)?>** and **<Colour (Colour) Is Touching (Colour)?>** code blocks can make for all kind of interesting gameplay integrations.

Interactive Story

Adding in some minigames or other interactive components could be a fun way to get the player more actively involved in challenges and puzzles. Maybe the player has to catch *x* number of things before they can proceed (or within a time limit)? Maybe they have to dodge and avoid something for an amount of time to survive a hazard? Maybe they need to time actions or reactions to clear a hurdle? Making the story even more interactive with gameplay elements can add a sense of danger or skill, so players have to "earn" the completion of the story. You could use an inventory display or map display similar to our Point-and-Click Adventure project to provide the player more information, especially for longer/larger stories. Other mechanics could include health, money, or other stats to control how the player can interact in the world; these could help add weight to the decisions the player is making, reward minigame performance, or act as a way to evaluate their performance at the end of the tale.

Snowball Fight

Our Snowball Fight was initially a fairly static affair; we knew the player's, opponent's, and snow wall's positions as well as the wind speed, and they remained the same until the bout ended. If you have an opponent and/or player that can move between throws, you could add a health meter system so more than one hit is necessary for victory. You could even have different damage rates based on where a hit lands, or for different weapon types. The snow wall could get different forms to allow for snow forts that could even include some ceiling coverage, and additional visible but non-interfering components to improve the looks of the game. You could try adding the Pen tool to trace shots or have a tracer special ammo so you can see and remember the flight path of your previous shots. The AI was very simple and didn't perform very well, so perhaps you could try making them a little smarter about things – especially not throwing the ball straight at the obstacle – and use that or other factors to add difficulty levels to the game. You could add multiple opponents or have player vs. player mode so two people can play! You could incorporate terrain so characters stand on a surface that can roll in valleys and hills, and the snowballs can collide dynamically with the surface instead of just ending at a particular y value.

Big Map Racing

To make it a competitive challenge, one could add a global leader board, though multiple leader boards to score each track separately would probably be a good idea. With tours or cups, players could race multiple tracks in sequence to create a single score for multiple challenges and simplify a leader board, or that could be an additional gameplay mode. You'll need to figure out how to handle unraced races – you don't want people scoring 0 for not racing when you want the fastest time! More ambitious additions could include a system for car design that would allow players to select different

Figure 10.4 The four advanced projects we cover in Book 3.

options of tires, engines, or chassis and have those choices tweak performance qualities (and possibly graphics) of the vehicle. A very challenging addition, but classic for racing games, would be a ghost car, an animation of another car racing around the track that doesn't interfere or interact with the player but shows them an idealized path/performance of the race.

Bar Charts and Data Files

Working with data lists is a wonderful skill to develop. This project was just a simple introduction, but there's so much more to explore and discover working with data structures. You can play around with data visualization in lots of fun and interesting ways. You could try making a line chart instead of bar chart by using the Pen tool. You could create colourization schemes based on data values. You could incorporate multiple columns of data by using multiple lists. With multiple lists, you can explore more chart types, such as a dot graph, or even try building a heat map if you're very ambitious. You can also try sorting data, see if you can take a randomized list, and sort it from smallest to biggest or vice versa!

Point-and-Click Adventure

There's one fairly significant change that's probably needed in this project besides any new additions. The *school door* sprite is so well camouflaged that players can easily miss it. Adding some kind of animation either on an idle cycle or on mouseover can be a great addition. A lot of point-and-click games have pop-ups or animations that react when the player's mouse hovers over them, or just randomly happen every so often. It is worth exploring the different styles you might use. To bring this particular story to life, you could try adding in other complicated animations, such as having characters move and animated along with the object animations. You could even add in some transition animations, like having the character take a bus to go from one location to another. Most commercial games in this genre have choices or gameplay change maps, scenes, and object availability within scenes. You could try incorporating some text entry for questions and answers between characters using some of what you learned in the Interactive Story project. Using what you learned in Bar Charts and Data Files, you could try using a list variable to track inventory instead of separate variables. You could also add in minigames or even health, money, or other stats to support new gameplay mechanics.

Platformer

The Platformer project had a lot of interesting terrains, hazards, and obstacles and works well as a template for students to build their own designs with, but we can still add a myriad of interesting features. A handy addition for both players and designers is a password/passcode system. By entering a password, players can jump to different levels, but most combinations won't work, so players must earn passwords by completing levels. This way, the player is helped by being able to return to the game to the furthest progress they had made, but also, the designer is helped by being able to jump to specific areas where they need something tested. The game could be more fully fleshed out by adding in a proper Loss screen or transitions between levels, which we left out for brevity. To extend the gameplay, there are many options for adding in hazards and obstacles – lava floors, fires, moving blades or saws, retracting spikes, opening and closing gates, trip hammers, falling spikes, projectile traps, lasers, or even moving enemies. We can also add in more beneficial terrain and tools – switch-operated bridges, teleporters, switches to turn off hazards, conveyors, or more. In the game we only had progression as a goal, but additional elements could be added, such as collectables or a timer, with which one could create a score system. One could also enhance the player character with animation cycles or the ability to have varying jump heights by button press duration. For the most ambitious change, one could try implementing the big map technique of Big Map Racing to allow for much larger levels, but one will have to think hard about how to implement all the components enlarged and tracked properly at scale.

Scrolling Shooter

Our final project was again a template for a well-established genre of games, so we've got plenty of easy inspiration for things to add. The most obvious are new enemy types. Adding in more enemies is a great way to keep things fresh, and tweaking the procedural generation to only introduce different enemy types as one progresses is a great edit to make. New enemies can have lots of new movement patterns: simple, repetitive patterns or the use of movement pattern system from the boss. Enemies could have different weapons/threats or different defences. For our player, new main weapons or special weapons and power-ups can offer all kinds of different fun and interesting experiences and flex our coding muscles. Having a power-up, load-out, or ship design system can let us explore lots of different qualities and abilities for the player, each offering different opportunities for play and design. You could add in menu selections for a load-out or ship design system, or possibly have in-game stores to buy new equipment. Beyond enemies we could add some hazards to the system, putting asteroids, mines, barriers, black holes, or other oddities into the battlefield that can interact with the player and change strategies. The game could be expanded with multiple distinct levels introducing new features or allow upgrading in between combat runs. One could add lore to the game more with story segments, intros, cutscenes, tips, or tutorials.

Extra Challenges

Not specific to any project, there are some common problems/features/challenges that you can always try to implement. These can require a bit of planning and consideration, so they can be great for pushing your own abilities or for keeping the advanced students busy.

1. Try adding a pause function and overlay. Can you push a button and pause the game? Often, it's more of a challenge than you might think!
2. Can you add some cutscenes? These can be great for conveying some aspect of a story, bringing a scene to life or prompting the player with some information and drive. Try starting or ending the game with an animated and music-, or sound-, enhanced cutscene.
3. Can you make it multiplayer? Can you add in an option for a second player to participate, either equally or as an assistant to the main protagonist? This can be very different things, depending on the project, but it's a fun way to not just share our projects more but change how they're experienced as well!
4. Can you add a tutorial within the project? It's important that our users know what to do; there's no better way to provide that information than right inside the project itself!
5. Cheat codes! A fun way to play some games is by including secret codes or easter eggs that players can discover that let them do silly, strange, or powerful things. See what you can come up with to change things around or add a new twist on the fun!

11

Troubleshooting Scratch

Perhaps nothing strikes fear into teachers told to integrate coding into their classroom more than the thought of dealing with bugs, errors, and computer trouble. Admittedly, coding can throw a lot of surprises at you. Earlier, we even said bugs are a part of the process. This might not be very reassuring talk, but just like coding itself, we can prepare ourselves for these eventualities. Here's some advice for the most common problems faced in the classroom when teaching with Scratch, in an attempt to arm you with the knowledge and practice to overcome most of the potential issues you'll face.

Demonstrating and working through bugs and errors with students is a fundamental part of working with technology. We want to show them resilience, determination, logic, and best practices for dealing with setbacks, and that's true with any subject, be it digital or analogue. By taking the time to make bugs and errors part of our teaching, we can use them to touch on a lot of key skills and make opportunity out of crisis. Working through bugs can build confidence and resilience, teach solution-finding practices, help practice analytical thinking, and provide for social learning opportunities with students helping each other with brainstorming, analysis, planning, and helpful solutions.

Site Issues

If you're using the website version of Scratch (as opposed to the offline downloaded and installed program version of Scratch), there can be some connectivity, browser, or website issues. That's without considering all the computer and connection issues – computers not booting, students unable to log in to their profiles, forgotten passwords, computers deciding now's the time to install updates, Wi-Fi not connecting, etc., which are all local issues with machines and networks. We are not dealing with those issues because they'll be specific to your school and school board, so hopefully you've been through all those issues with your local tech support. Instead, let's focus on issues particular to Scratch, assuming you've successfully reached the website.

Problem: The project is freezing/not responsive.
Possible Solutions:

1. Refresh the website (click the Refresh button in the browser).
2. Does the project use clones? If so, check if too many clones are being generated (over 300 can overload the website). Clones, repeats, or My Blocks generating clones is a common cause of this issue.
3. Close and restart the browser.
4. Reset the Internet connection.
5. Give it time (website may be temporarily down or overloaded).

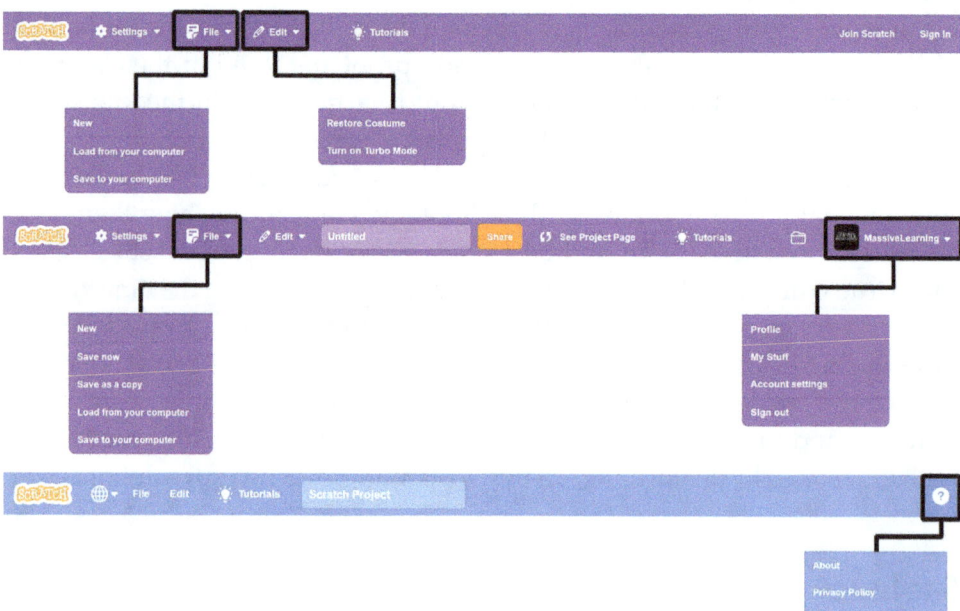

Figure 11.1 The status bar's File options when not logged in (a guest) or when logged in. Cloud saves and sharing are not available unless a user is logged in!

Troubleshooting Scratch ◆ 195

Problem: Can't remix projects.
Solution: Log in to a Scratch account.

Problem: Can't autosave or save.
Solution: Log in to a Scratch account.

Problem: Can't name projects.
Solution: Log in to a Scratch account.

Problem: Can't access/save to/see the Backpack.
Solution: Log in to a Scratch account.

Problem: Can't access previous projects.
Solution: Log in to a Scratch account.

Problem: Can't find/see the project that was sent.
Solution: Project owner must ensure the project is shared.

Problem: Link doesn't work.
Solution: Project owner must ensure the project is shared.

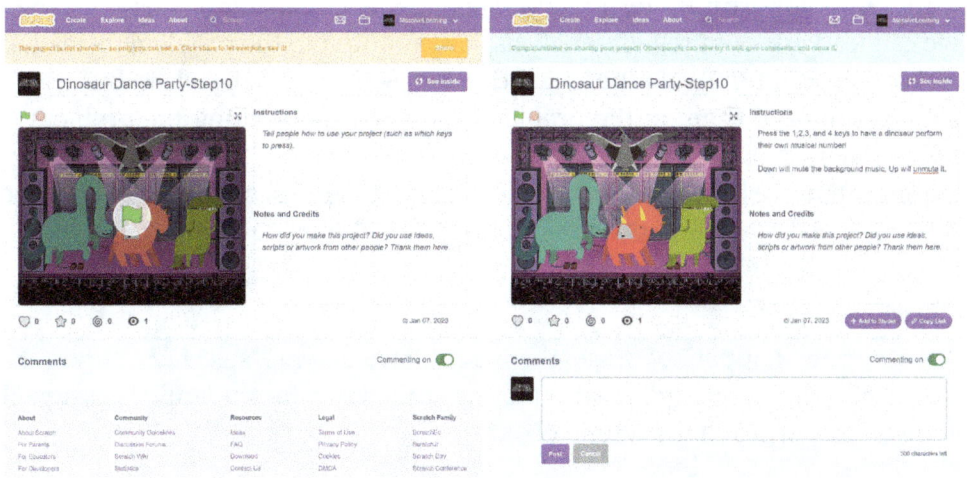

Figure 11.2 You don't have to share what you create, but it does help build the community. For education, it can allow teachers to share starter projects, or for students to do peer review or submit projects. Here's a project before and after sharing.

Coding Issues

Most errors and issues you face will be coding problems. Computers are very literal in their understanding, and this is something most humans are not very good at. Our brains are built to predict, imagine, and leap to conclusions to act quickly and save time and energy. We often make assumptions or leaps that computers can't and, making our code, end up doing something we don't expect as a consequence of it. There are many forms of error and many reasons, but the fundamental process of thinking being different between humans and computers will likely always cause friction and bugs. The most irritating truth of bugs is, no matter how unwanted the behaviour is, the computer is only doing what we tell it to do.

While the site issues we listed earlier are relatively mechanical – and you probably already deal with those kinds of issues regularly – coding issues are a whole new can of worms. Most importantly, we want to prepare ourselves for how we think and deal with bugs and errors before we even start fixing them. A bug or error is, like in any other subject, an opportunity for reflection. When a student says something is going wrong, we don't want to rush to fix it for them. Can they identify the problem? Can they enunciate it? Can they solve it themselves? Can they eliminate some of the possible causes? Our first goal should be to challenge the student to overcome the obstacle, or at least give it their best shot. If they engage and try those things, we still don't have to leap in ourselves to fix it. We can then challenge the class to think about the problem, discuss it, and suggest solutions. This helps take the problem and turn it into opportunity for practice and class engagement. If that fails, then we can lead our own review, analysis, and solution to the problem instead of just solving it for that one student and ignoring the rest of the class. By having already engaged them in the problem, our time spent fixing the problem is helping inform all of them.

Now that we have some better idea of how to approach bugs, what are we dealing with? There are four basic areas where things go wrong.

The Wrong Object

These errors happen because the wrong object was selected. Either it's displaying the wrong information to the student or the student has assigned code or properties to the wrong object.

Problem: "I can't find that code block" or "That code block isn't there" – the list of code blocks is different than expected.

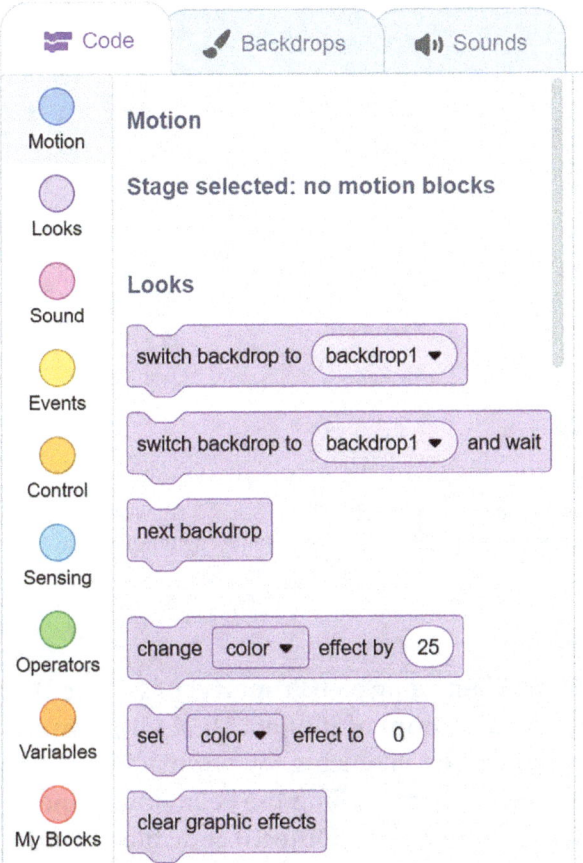

Figure 11.3 When the stage is selected, code blocks that don't apply (such as all motion code blocks) don't appear.

Solution: The student has selected the stage instead of a sprite. The stage doesn't have a number of the code blocks available to it since it can't move or do a number of other things that sprites can. Make sure the student selects the correct sprite first.

Problem: The wrong object acts or reacts.
Solution: Code was placed in the wrong object. Check the object that did act/react and the one that didn't, and you'll likely see some code assigned to the "wrong" object.
Alternate Solution: The wrong broadcast event might have been called, making an unexpected reaction happen. If the behaviour you saw is supposed to happen at a different time, that would indicate a wrong event instead of a wrong object issue.

Figure 11.4 You must be logged in to access the Backpack. ❶ shows none. ❷ shows the minimized Backpack. ❸ shows the Backpack maximized with some costumes, scripts, and sprites already added.

Alternate Solution: When using a **((Property) of (Object))** driven behaviour like **[Move to X: ((X Position) of (ObjectA)) Y: (0)]**, if you selected the wrong object, you could be referencing the wrong object's properties and making it look like one object is calling the behaviour rather than the other. Double-check any **((Property) of (Object))** code blocks and see if they are correct.

General Tips
Transferring code blocks. Remember all the ways to copy and transfer code blocks (drag, copy + paste, the Backpack). You don't have to rebuild things from zero. Save time by copying code that's incorrectly in one object over to the one that actually needs it to save time (but remember to delete the incorrect code after copying!).

   Specify Objects. While instructing, be sure to specify what object you're coding for and when you switch objects. Repeating yourself to the class can be very useful for avoiding these mistakes.

The Wrong Block
There are a number of blocks whose similar nature can easily trip up new coders. These errors can also be hard for people to spot because of our habit to read what we expect rather than what's there. Knowing the likely culprits, you can look for them easier and quicker, as well as predicting the issue and

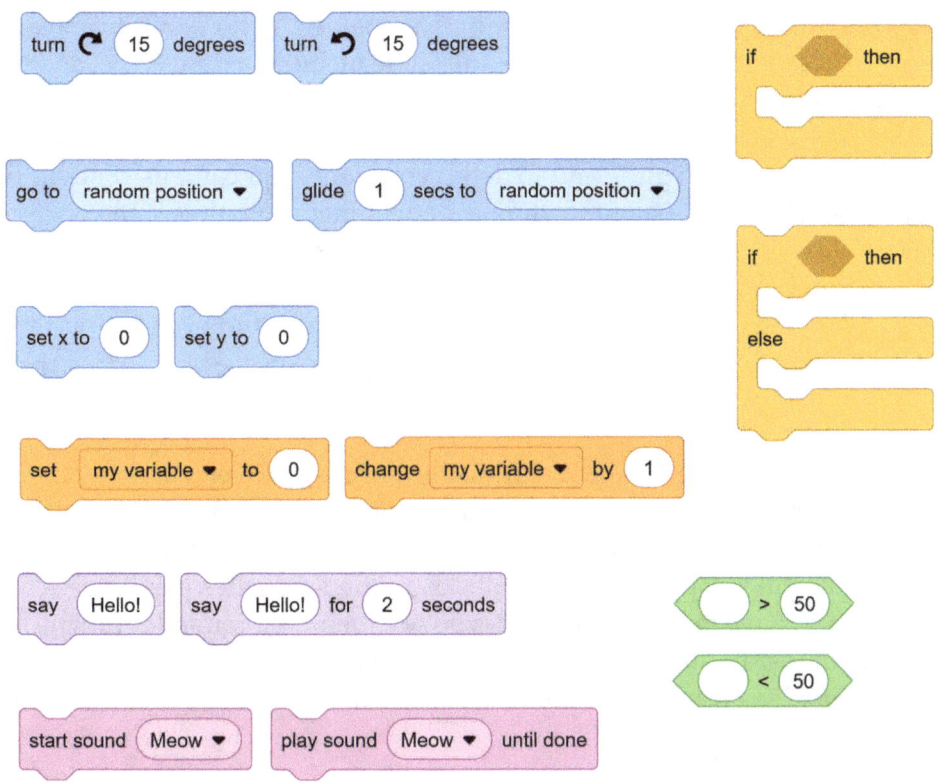

Figure 11.5 The most likely code blocks students will mix up. If your project uses one of these, expect somebody to grab its twin!

making sure you clarify the difference when they come up in instructed learning.

Confused Pair: Left vs. Right
The turn code blocks for left and right turns can easily be confused, especially for younger students still uncertain with their directions. It's unlikely to cause serious issues, but if something is veering off in an unexpected direction and you used a right- or left-turn code block, check if there was a switch. Sometimes, seeing the difference can actually be fun and interesting, like in our Pen Tool Fun project.

Confused Pair: Go To vs. Glide To
Both of these code blocks can send a sprite to a specific X/Y or sprite location, but they work differently. •**[Go To]** automatically teleports a sprite instantly, changing its X and Y properties. •**[Glide]** works by changing the X and Y properties over a set amount of time. Kids love glide because it animates, and

they can see the sprite moving (gliding) to the destination, but it can cause some issues if used in place of •[Go To]. Glide includes a time value, and this acts as a hold or delay on the code. Until the glide is complete, the computer won't move past that code block, which can cause some issues. If you want a sprite to move with glide and to detect collisions, you can't use both in the same stack, because while the glide is happening, it won't run the collision detection code because it hangs on that single timed code block. •[Go To] can cause its own problems, as it changes instantly, so sprites look like they're teleporting and might be too sudden a change; you could even teleport past an obstacle without ever colliding with it like you expected.

Confused Pair: X vs. Y

It can take a while for X- and Y-coordinates to sink in. Even adults who have learned them can forget when they haven't been using them. When things move left/right or up/down unexpectedly, you can always double-check any X- or Y-affecting or powered code blocks. In the Properties panel, the X and Y properties have appropriate arrows beside them as a reminder of which is which, but there's also plenty of mnemonics you can introduce your class to (X is a cross/X is across, or Y points down, etc.).

Confused Pair: Set vs. Change

There's a lot of features with set and change versions of code blocks: X/Y positions, size, graphic effects, volume, sound effects, •variables, and pen properties. Until students get used to working with variables, it's very easy getting confused between the two. Set makes a property/variable an exact number, regardless of what it was (absolute). Change adjusts a property/variable from its current position (relative). This difference between absolute and relative changes can be hard to describe, but projects can be a great way to visualize the difference, and it can be well worth taking the time to illustrate the difference with a visual example like position or size.

Confused Pair: Say/Think vs. Say/Think For

All four of these code blocks allow displaying some text on the screen. The •[Say] code blocks use a speech bubble, while the •[Think] code blocks use a thought bubble. The difference is "For", where a value is introduced, allowing a time scale that makes the displayed text disappear when it is up. Without "For", the text is permanent and only disappears with a •[Hide] or new •[Say]/•[Think] code block. The permanent versions are good for non-action, non-animating scenes, information displays rather than reactions or conversations. One needs to be more involved in ensuring they disappear when you need them to. "For" versions are great for fast-paced programs, where you want to maintain some action and activity, and you can set and forget them since they'll disappear on their own. Be careful the information displayed doesn't disappear too quickly for slower-reading users.

Confused Pair: Play Sound vs. Start Sound

Similar to •[Go To] vs •[Glide], the difference between these two blocks is time. A •[Play (*sound*) Until Done] code block plays a sound effect, but it holds the program on that code block until the sound effect has finished playing. A •[Start Sound (*sound*)] code block begins playing a sound effect but doesn't hold the program (other things can happen while the sound is playing), and code execution continues on without any delay. So •[Play (*sound*) Until Done] is useful for things like alerts or dialogue, where you want to pause things until it completes, whereas •[Start Sound (*sound*)] is great for sound effects and other incidental sounds that may even stack up and simultaneously play while other things happen.

Confused Pair: If vs. If/Else

This pair confusion won't tend to last long with the block structures limiting coding options, but I thought it is worth mentioning. It can be hard for younger kids to see an •[If <True> Then] code block and realize it isn't the one they want, so you need to be very clear about selecting the right one with younger audiences. The difference is, of course, the "Else" clause. An •[If] only tests a condition and allows code if the condition is found true. In an •[If/Else], the condition test determines which of the two clauses of contained code is used, the first if the condition is true, the second if not.

Confused Pair: > vs. <

This will be no surprise to teachers, coding experience or not. The •Operators blocks for numeric comparison can be confusing to younger students. Again, the key is clear communication beforehand to ensure the correct selection in the first place. We can use all the math class mnemonics to try to get the difference clear with our students. The nice thing about code is that with the computer responding immediately to the code, students can test and see the results on their own.

The Wrong Order

Even with the right sprite and the right code blocks, things can go wrong. Our third category of errors are "wrong order" issues, where code blocks aren't in the right sequence to execute as desired. This is where we start getting into more complicated issues. Thanks to Scratch's code block system with distinct shapes and colours for code blocks, a number of issues are much more visible and easily avoided. This visual feedback doesn't just help students follow and build things but can also help avoid or correct bugs.

Simple Sequences Code Flow

The most basic errors in this category are simply putting a code block above or below where it belongs. In many cases, the exact order doesn't matter. For example, in giving multiple variable values to initiate a program or setting multiple properties to represent a new state. In these cases, the exact order rarely matters; the properties are all set in a single frame of animation and don't interfere or influence each other. But in other cases, the order is extremely important. Take for example a sprite given a •[Move (100) Steps] then •[Turn Right (45) Degrees] versus a sprite given a •[Turn Right (45) Degrees] then •[Move (100) Steps] instructions. They'll end up facing the same direction, but almost as far from each other as they are from their starting positions.

Control Structures and Code Flow

As we introduce control structures like •If statements or •Broadcast events, we change how the code flows. Thinking of code through flow charts or process systems, errors can occur because we took a wrong turn in our chart. This could happen because we called the wrong •Broadcast event or were listening to the wrong one, triggering the wrong code to execute out of sequence. It could occur because we changed a control like a game state that triggered an •If statement at the wrong time and changed the expected sequence of events or a property at the wrong time. •Control structures are very powerful and handy, but a "typo" of grabbing the wrong block or value can have us swerve off course in the flow of processes.

Logic Clauses and Chains

Learning to use logic can be a challenging task; there are lots of errors to be made when including or excluding conditions in code. It can be hard to know whether to write some conditions inclusive or exclusive, whether to use an "If", or an "If/Else", or which two clauses to put your code in. If something strange is happening, always check the logic conditions; it's surprisingly easy to end up with the reverse of what you intended because of our love of leaping to conclusions, leaving the cold mechanical reasoning of computers in the dust. One very easy issue to fall into is chaining logic conditions – when you put multiple logic operators together for a single test. **<<<A>or <B>> and <C>>** will give different results than **<<A> or <<B> and <C>>>**. It can be very hard to spot this kind of error, so know to look for it!

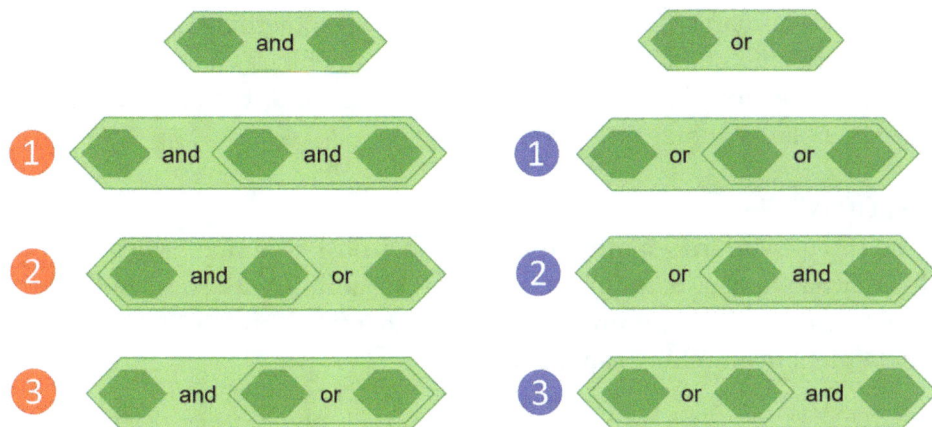

Figure 11.6 Combining logic blocks can easily become confusing. It's important to remember that how they are connected may affect the results. ❶ Ands can be combined and simply add to the necessary conditions to meet. ❶ Ors can be added to add additional options to meet. When mixed, the outer code block is dominant. So in ❷ you must have both the first conditions or have the third. In ❸ you must have the first and one of either of the second and third conditions. In ❷ you can have the first condition or both of the second and third. In ❸ you must have one of the first or second conditions plus the third.

Concurrency and Race Conditions

We can run into other problems in the flow of our code processes separate from those previously mentioned. Sometimes issues happen when we need to handle multiple processes that happen concurrently, or with (near) simultaneous execution of code. This can cause specific problems with order of execution. If any properties or variables are affected in one process that affects the other, the order of operations could drastically change the results. It can be challenging to make simultaneous systems work without improperly interfering with each other. When we don't know which process will complete first and they can influence each other, we call that a "race condition". The faster process "wins" and makes changes that could affect the other. Sometimes, the order can be predicted, sometimes not, so these issues can be very tricky. We can try to put code into single stacks or split it into multiple stacks to simplify processes to a linear run or to ensure concurrent execution as needed. Single stacks make the flow much easier to analyze and predict, but sometimes concurrency is exactly what you need. As we mentioned earlier, take for example the •[Glide] code blocks: if an object is gliding and you need it to be clicked or collided with, you'll need to put the collision or clicking code in a concurrent process because other code in that stack won't execute until the glide is completed. Delay blocks may or may not help simplify these issues. Concurrency can be a blessing and a curse and usually very hard for the human brain to predict accurately. If something isn't happening, or it's happening at the wrong time, check for any block with a time value or time-sensitive word like "until", "for", or "done" and think about how they might be holding things up.

Other Errors

These three categories merely represent the most common issues a coding educator has to deal with. In this category, we'll deal with some of the other less-expected issues that can happen in Scratch that don't necessarily fit into a convenient overarching category.

Layers

Some of the simplest issues come from layering. Every graphic on the screen is printed in layers, each object laid on top of the objects below it, all on top of the backdrop. At times, this layering may cause issues. Large objects may completely obscure smaller objects underneath them. You may need to control the depth of objects at times, but the •[Go To [Front]], •[Go To [Back]], •[Go [Back] (1) Layer], and •[Go [Forward] (1) Layer] code blocks can be used to solve these issues.

Visibility

There are a number of issues you may run into dealing with object visibility. The first and most obvious is having objects disappearing. If you ever •[Hide] an object, you will need to have a •[Show] code block somewhere to make it visible, or you'll never see it again after it first disappears. While invisible, or hidden, a sprite has limited functionality. A hidden object cannot •[Say] or •[Think]. A hidden object also doesn't collide with other sprites.

Colour Selection

Scratch uses a graphical technique called anti-aliasing. This is a process that blends (and changes) colours to smooth out lines to appear more visually appealing and avoid the jaggedness that pixel designs and displays can cause. This looks great, but it can lead to a few tricky issues. If you are using any colour collisions, watch out for anti-aliasing blending the colours you think are involved. If you want to use a colour and use the colour picker, you may see that there are many more colours than you expected in the zoomed view of the screen. Any colour near a line or change of colours may end up being blended by anti-aliasing. This means that colour collisions looking for that colour won't find it because it changed to a different colour. This is especially likely to happen with sprites that are being displayed at less than 100% size or at anything but 0% ghost effect, which blends them with whatever is behind them to create the transparency effect. Colour collisions are great, but if you're having unexpected results, it may be because of these issues.

Clones vs. Originals

As we mentioned in our projects, clones can be tricky. Whenever you're using clones, remember that there's a difference between the original sprite and any clones you make. The original sprite will never receive the ●[When I Start As A Clone] event, and the clones will never receive the ●[When ▷ Clicked] event. Don't expect the original to act like a clone, and don't expect clones to act like the original. It can be hard to get used to it, but once you do, it is a very powerful system to work with. Also, remember that clones always start with the same properties as the original. If you don't see clones, is it because you had the original ●[Hide]? Unless you ●[Show] the clones, they'll remain hidden.

Wrong Concepts

The hardest of bugs is when we simply had the wrong conception of a system or method. Our brains ran off with an idea that wasn't true, doesn't work that way, or otherwise doesn't line up with reality. There's no predicting this and no preventing it. Sometimes we have ideas that don't work. We want to try to make them work and account for any bugs in the system, but sometimes the way we conceived something is simply not how it will actually work. Maybe we didn't account for some factor that makes it unable to work, or it is simply inefficient or impractical. It happens. Even to the pros. The way to address this issue is always approaching coding with a flexible mindset. There are many ways to address an issue, and sometimes we need to explore and try some options in order to really understand the problem or our tools in order to come up with better ideas. Learning can sometimes mean failing. Learning that something doesn't work, or work in that way, or work in this condition, is still learning. Learning those limitations and finding ways to work around them is perhaps the greatest part of learning coding.

Backup Plans

Having been a coding educator for years, I've seen just about everything go wrong. The worst issues are large-scale technological issues out of your control: the Internet is not working, power outages, websites go down, etc. Most of the time, educators can simply switch tracks and work on something not so technologically dependent, but for those of us that are coding specialists, how can we deal with the wrenches that sometimes fall into the gears? Here are some ideas about how to work around dealing with major technological hurdles and still deliver some educational opportunities.

Offline Scratch

One of the most impressive things about Scratch is that they have both the online instant access website and a downloadable offline version. If you have issues with spotty or slow connections, you can download it and install as an app that requires no Internet connection to work. This can be a wonderful tool for rural and remote communities, to ensure lack of Internet doesn't lead to lack of learning access.

Some educators may actually prefer the offline version of Scratch. By having the app, students won't be distracted by the shared content on the Scratch website. While I think the inspirational factor of the sharing platform side of Scratch is a great benefit to students, some may benefit from less distractions.

Another factor going for the offline version of Scratch is that it means students will save and sort their projects in their student profiles. This does mean the classroom organizing and viewing options aren't available but for prolific creators, but the ability to sort projects into folders in a way that personal accounts don't allow for can be a big benefit as dozens or even hundreds of projects pile up.

Most school board computers will have program installation locked down (and for good reason), so if you want to work with offline Scratch, you'll need to have the tech department install it on the student computers. The nice thing is, this can mean having someone else do the work and possibly automate the process, so it shouldn't be any additional hassle for teachers to have this capability in their classroom.

Pseudo-Coding

Let's be clear: "pen-and-paper" coding is absolutely no replacement for working with computers, but learning to do pseudo-coding can be a powerful and useful skill to add to the mix of tech skills we develop in our students. Pseudo-coding is when a coder writes out their design, not using the exact code word for word, but in a shorthand to give the broad strokes of their plan on how to achieve or organize things. It's more about planning than syntax and details.

Pseudo-coding can look different for different people, as each person makes plans or notes in their own way. In general, pseudo-code will help define the components of a project – such as the objects, backdrops, costumes, sounds, writing needed, as well as the processes, the events, and the stacks that will make things happen. Pseudo-coding can be as verbose or as brief as needed; plans can be quick sketches or fleshed out fully to the point of actual code, but often it consists of shorthand sentences, like "If near [objective] then change to warning state". Maybe they don't know what they want the warning state to be, or maybe they do, or maybe they have some idea about a feature like "Repeat 10 {size +10%, wait, size -10%}" as a note to make something pulse and a method to achieve it. You'll note the exact name of code blocks might not be listed, but the intent is clear.

In addition to more written pseudo-code, there can be very different methods for keeping these notes, as the creative process works differently for different people. Some may use more text document format, others may want to use spreadsheets to list ideas, some may prefer physical whiteboards, others might use presentation software to make things more visual, and flow charts could be used, mind maps, or even wikis for very large projects. No perfect system exists for pseudo-coding, creative designing, or visioning, since everyone and every project is different. Explore and experiment with different methods and tools, see what you like, and help your students explore all the ways their ideas can be brought to life. There are so many wonderful ways to be creative, and such wonderful things to create. You and your students have beautiful, thoughtful, meaningful, and fun treasures hidden inside them; by discovering and sharing the tools and opportunities to bring them to light, we all benefit. We hope this book has helped you learn, grow, and prepare to be a guide on that process and help usher in a beautiful era for human creativity and potential.

12

Final Thoughts

With that we complete not just this book but this series. I'd like to thank you for taking the leap and giving coding education a fair shot, and for the honour of being your guide to get started. In this last book, we've searched out the last remaining functionality of Scratch we hadn't yet put to use. We've pushed up against the ceiling of its capabilities, seeing how our ideas have pressed up against its protection protocols, or the size and complexity of our projects are beginning to strain the use or comfort of working with the editor and code blocks. You can never complete learning coding any more than you can finish learning art. We've developed our skills and understanding to a level of mastery, though, able to engage and use Scratch to its full potential.

Our purpose for the series wasn't limited to teaching ourselves how to code but to prepare ourselves for using this knowledge to teach our students. While we haven't focused on pedagogy and practice (we're saving that for another book), your technical skill in Scratch was our primary goal. We sought to make sure you'll have the confidence working with Scratch and coding, from which you can build a pedagogical practice. By incorporating Scratch into your teaching practice, you'll not just check the box of whatever coding requirements your school district requires but also be able to build a positive, collaborative community of creators in your classroom. We aren't going to just teach coding with Scratch; we're going to be able to use coding in Scratch to explore any subject and idea. We've unlocked a tool for creation and exploration to expand our classroom's and the student's capabilities. We'll never know all the answers, or code without bugs, but we'll be able to work together

Figure 12.1 The three books in *The Teacher's Guide to Scratch* series.

and explore, collaborating to create new ideas and to work through problems. Ideally, we teach so that the student can exceed their master.

This advanced level of coding is a great preparation for text coding. Scratch introduces the same structures and concepts, as well as a number of terms, that will help students move into text coding with some familiarity. With advanced Scratch, we've started working with some of the classical CS training concepts and methods, helping prepare students for CS10/20/30 classes so they can, if they choose, move seamlessly into specialized education. When they're ready, they'll be able to move on from Scratch to professional languages and platforms.

We don't view Scratch education as the means of directing students toward CS courses or programming careers, though. Coding is for everyone, even those with no interest in pursuing it past Scratch. CS can remain a specialist science, but Scratch has allowed us to set a base level of coding education for all students – accessible, engaging, meaningful, and purposeful. Scratch will allow us to revolutionize education to the digital era, ensuring a technical understanding to underpin the goal of every student being an aware and engaged digital citizen. Students will be able to see themselves as digital creators, not just digital consumers, giving them agency in the digital world.

Coding educated students will have a better appreciation of the underlying technology our society is built on and operated by. This will be critical as we move toward further automation and artificial intelligence. We will need informed and engaged citizens to ensure our politics is informed and guided by a citizenry (and comprised of candidates) that have a good grasp of the realities of technology, which we have been sorely lacking thus far. The future of democracy, and its ability to handle the social upheaval of artificial intelligence, depends on our work to provide our students this education.

So thanks to a little guidance, and a lot of playing around, you've mastered Scratch! You now have an incredible new tool at your disposal. You can use it to create, inspire, illustrate ideas, inform, and challenge your students. By incorporating a creative tool capable of data processing and modelling, you'll be able to provide amazing opportunities for your students. You can use it to demonstrate any number of concepts in any subject. Most uniquely, as you dive deeper with your students, you'll be able to explore logic, computer science, and information theory. Through this tool you can reveal some of the most fundamental principles of the universe. Through playful and creative exploration, they'll be able to be introduced to, grasp, explore, and master all kinds of concepts. By incorporating coding through Scratch in your classroom, you'll be providing the means to explore, understand, and create the world of tomorrow for your students. Congratulations!

Glossary

⚐ – This icon is used to represent the green flag that is used as a start button for Scratch projects. It can refer to the start button in the Scratch editor above the Stage Window, or it can by in a •[When ⚐ Clicked], which is the event that runs when a user clicks the green flag button.

✏ – This is the icon we use to represent the Pen tool extension in Scratch. The Pen extension, when added to a project, adds an additional category to the Code Block Library; instead of a coloured dot like other categories, it has a pen icon. When you see this icon, it means the following code block is in the Pen category.

.sb3 – This is the file extension for Scratch 3 saved files. Earlier versions of Scratch used. sb1 and. sb2. You can *"save to your computer"* and download an. sb3 file of a project and *"upload from your computer"* to load a file into Scratch.

Abstraction – This is one of many terms used to define computational thinking. It is the process of breaking a problem or goal into simplified and standardized components so simple and reusable processes can be developed to address those components. See also *computational thinking*.

Algorithm – An *algorithm* is simply a defined process to find a solution or achieve a goal. Math formulae are examples of simple algorithms. Recipes are another form of algorithm – a clear set of delineated instructions leading to a goal: food! Computer programs may be an algorithm or contain one or more algorithms.

Algorithm Building – This is one of many terms used to define *computational thinking*. It involves creating defined processes that, if followed, solve, or help solve, a portion of a defined problem or work toward a particular goal.

Argument – In computing, an *argument* is a piece of data passed to a function. Arguments are used by the function to evaluate a procedure or subroutine. They may also be referred to as *values* or *parameters*, though each has its own similar but distinct specific definition.

Art Workspace – In Scratch, it's the central portion of the editor when in the Costumes tab or Backdrops tab. It is comprised of the art tools and the canvas.

Asset – In general computing, an *asset* is something, not code, that a program uses to complete its function. This commonly includes graphics, sound files, 3D models, or data files.

Avatar – A person's representation on a digital platform. An icon, sprite, or even 3D model may serve as an avatar, depending on the context.

Backdrop – A graphic displayed by the stage in Scratch. The stage can have multiple backdrops, but only one can be displayed at any given time.

Backdrop List – When viewing the Backdrop tab in Scratch, the Backdrop List is displayed on the left-hand side of the screen. It shows all the current backdrops assigned to the stage that could be displayed and highlights the one currently selected.

Backdrop Tab – In Scratch, when the stage is selected, the second tab in the upper left-hand corner of the editor becomes the Backdrop tab. If clicked on, it provides access to an art workspace to create or edit backdrops and shows the backdrop list on the left hand side.

Backpack – In Scratch, if a user logs into their Scratch account, they can access the Backpack at the bottom of the editor. A small button labelled "Backpack" will show when collapsed, and clicking it expands to show a single row of thumbnails of what is currently in the Backpack. Users can drag scripts, sprites, backdrops, or costumes to the Backpack, where they will be saved and be able to be copied out from into any project the user opens. Items can be added to the Backpack by clicking and dragging them onto it. They can be deleted by right-clicking on them and choosing Delete on the pop-up.

Boolean Block – In Scratch, these code blocks are distinguished by their sharp, poky sides. They are used to provide arguments for some •Control blocks, such as •[If <condition> Then]. They are used to perform an evaluation that will return true or false, also known as a Boolean result. Most Boolean blocks are in the •Operators category, but there are also Sensing category Boolean blocks and others. The •<(0)=(0)> code block is an example of a Boolean block.

Broadcast – In Scratch, *broadcasts* can refer to a few different •Event code blocks or to the unique named signals called Messages that are broadcast or received through them. Broadcasts are triggers that can allow code to trigger other code, including across multiple sprites or the stage.

C Block – In Scratch, numerous code blocks are shaped like a C or E, with gaps inside them that can fit other code blocks. These are called C blocks. The •[If <condition> Then], •[If <Condition> Then {} Else {}] are examples of C blocks. Code that is placed in the gaps of a C block is considered "nested" inside it in computer science terms. C blocks will exert some control on how the code placed inside them runs. Most C blocks are in the •Control category.

Canvas – In Scratch, the *canvas* is part of the art workspace where you can actually draw and create images. The canvas has a centre point indicator and a line marking the 480 × 360 pixel area that will cover the entire Stage

Window, but it extends past this area. In general computing, the term *canvas* can apply to anywhere you can draw or show images, but its exact use may be more precise depending on the platform/context.

Cap Block – In Scratch, any code block that ends a script (it does not have a bottom connector, so no code can go below it) is a cap block. The most common cap block is **[Stop [All]]** in the Control category. The **[Forever {}]** code block is both a C block and a cap block because it repeats its nested code blocks endlessly and nothing can ever come after it.

Character – In computers, the term *character* can refer to either a typographic symbol, like a letter, number, or punctuation mark, or to a fictional being in a game, interactive novel, or other digital media. A character in a game could be a player character (PC), one that is controlled by a human player, or a non-player character (NPC) that is controlled by the computer. See also *player*.

Clone – In Scratch, a *clone* is a copy of a sprite. They function just like the sprite they are copying, but they never run the **[When ▷ Clicked]** event since they are created while the project is running and therefor do not exist when that event is triggered. They do, however, get access to the **[When I Start As A Clone]** event, which the original sprite they are duplicating will never be affected by. Clones are a very powerful technique to use in Scratch for populating a busy world or creating special effects, but Scratch is likely to crash if you create more than 300 clones at a time. Clones are dealt with through a trio of code blocks in the Control category.

Cloud – In general computing, *cloud* systems are massively scaled and redundancy-protected systems to handle user data or even run programming. The online version of Scratch is a cloud system, saving user accounts and data on massive web servers that users access through the Internet.

Code Block – A compartmentalized instruction for a computer system and fundamental unit of block coding. Each code block has a specific function that it makes the computer perform. Connected together in the right order, they can create algorithms and programs, giving full complex instructions for the computer to fulfil a given series of tasks. Each code block's shape and colour will provide users with details about how they connect and what its function might be, or how to find it in Scratch's editor. Keep in mind: code blocks are not unique to Scratch, and there are many coding websites and programs that have adopted using them as the means to program.

Code Block Library – In Scratch, when the Code tab is selected, the Code Block Library is displayed along the left-hand side of the editor. It lists all the code blocks available to the user to use in their project. They can scroll up or down through the list, or along the very left-hand side of the editor, they can click on any of the categories (Motion, Looks, Sounds,

•Events, •Control, •Sensing, •Operators, •Variables, or •My Blocks) of code blocks to jump to that category. The extensions add new categories and code blocks to the Code Block Library for that project.

Code Tab – In Scratch, this tab in the upper left-hand corner of the editor gives users access to the coding workspace in the middle of the editor as well as the Code Library along the left-hand side of the editor. In the Code tab, users can access the code blocks and create code for sprites and the stage. To make art, the Costumes tab or Backdrops tab are selected. To make sound, there's the Sounds tab.

Coding Workspace – In Scratch, when the Code tab is selected, the centre of the editor provides a workspace to do coding. Here, the user can arrange code blocks they drag from the Code Block Library to the left or by copying them in. The coding workspace is infinite, and users can click on blank areas of the workspace to drag it and change their view, creating more room as their projects grow. The code shown in the coding workspace is specific to the current selected sprite or stage. In the upper right-hand corner of the coding workspace, there will be a ghost image or watermark of the sprite or stage being coded for.

Collision Mask – In computing, a collision mask is a graphic that determines the positions that an object occupies and can collide or overlap with other objects. It may or may not match the size and shape of the object's display graphic. This method allows objects to have different visual looks compared to their collision behaviour, so in games objects can have artistic perspective that doesn't cause collisions despite literal graphic overlap.

Colour – In general, *colour* for digital systems is defined by three numbers assigned to red, green, and blue, or RGB. These control how bright each component light, affecting how a pixel on a digital screen looks, is turned on. If all three numbers are 0, the pixel is black, and if all three are maximum (generally 255) the pixel is white. Different combinations of values on each primary colour create different colours. In Scratch, *colour* refers to an exact output colour, what a pixel is coloured, but also refers to the hue of the final output colour, which is also affected by its saturation (how vivid or pale the output colour should be) and brightness (how bright or dark the output colour should be). It can also refer to an argument for the •Looks •[Change [Graphical Effect] By (#)] or •[Set [Graphical Effect] By (#)] code blocks that are used to shift the colouring of a sprite's costume, as opposed to the other possible graphical effects (like ghost or pixelate).

Colour Panel – The colour panel is a pop-up in Scratch that allows users to select a colour or fill pattern. It is used in numerous places wherever colour selecting is needed. In some places, some features of the colour panel are not available, depending on the context. Tips on working with the colour panel are in Chapter 4, "Scratch Basics".

Colour Picker – The colour picker is a tool accessed through the colour panel in Scratch represented by an eyedropper icon. It allows users to select a colour already in use, either in the canvas or the Stage Window, depending on the context. It will provide a zoomed-in view of an area and allow a single pixel to be clicked to provide that colour to the colour panel.

Combo – An unofficial term in block coding for whenever a code block is combined with another, especially when a Boolean or reporter code block, or code blocks, are placed into another code block. We also use the term *combo* to refer to some commonly connected code blocks, such as •[When ▷ Clicked] followed by a •[Forever {}].

Comment – In general computing, a *comment* is some writing in a program that is not seen by the user or interpreted by the computer but rather simply acts to inform anyone reading the code of something. Commenting code is a good practice that can help make it more readable by providing titles, declarations, explanations, and footnotes to the programming. In Scratch, comments can be added to either the coding workspace (by right-clicking on it) or to a specific code block (by right-clicking on it). Then, a sticky note graphic will appear. It can be moved, scaled, and typed into to provide a comment.

Computational Thinking – In general education, the concept of computational thinking helps define a system for problem-solving. It is comprised of decomposition, abstraction, algorithm building, pattern finding, modelling and simulation, and evaluation. It provides practical methods to solve problems through both analysis, planning, and execution. See also *decomposition*, *abstraction*, *algorithm building*, *pattern finding*, *modelling and simulation*, and *evaluation*

Conditional – A *conditional* is a system of evaluation in computing. One or more metrics with one or more evaluations made of it returns a Boolean signalling if the condition is either true or false. For example, to check the weather in code could include •[If •<•(Raining) = (True)> Then]. Conditionals can also be numerical, as in •<•(Temperature) > (32)> or any other evaluation or comparison you might need on your project. In Scratch, conditionals are used with •Control category code blocks to make code run only if certain conditions are met so projects can be reactive and dynamic.

Costume – In Scratch, this is the graphic displayed to represent a sprite in the Stage Window.

Costume Library – In Scratch, if the user is in the Costumes tab, they can click the Add a Costume button to add a costume from the Costume Library. This lists all the graphics available by the default user.

Costume List – In Scratch, if the user is in the Costumes tab, all the costumes currently assigned to the selected sprite are shown on the left-hand side of the editor. The user can scroll up or down through the list. The selected costume displays in the art workspace. The user can click on any thumbnail in the Costume List to select it, and if desired, they can drag it up or down to change the order of the costumes in the Costume List.

Costumes Tab – In Scratch, the Costumes tab is available in the upper left-hand corner of the editor if the user has selected a sprite; otherwise, the stage selected and the Backdrops Tab will show in its place. Clicking on the Costumes tab allows the user to access the Costumes List and the art workspace to edit or create new costumes for the sprite they have selected.

Creator – In general, a *creator* is anyone using a program, or skill, to create something. In Scratch, it can mean the person using Scratch, the user of the editor, also known as a "Scratcher", or the person who created a project or asset that is being remixed. Because projects can be remixes of remixes or include assets from other projects or from outside Scratch, what exact relation a creator has to a project can be complex.

Decomposition – In education, *decomposition* is one of the components of computational thinking. It is based on breaking up a problem into distinct components to make smaller and easier tasks to deal with. This is useful in all problem-solving but particularly handy in coding because we can divide complex needs into discrete goals that can be easier to analyze and build solutions for, leading toward the end goal.

Define – One of the steps in the design thinking product design process. It comes after the empathise step and before the ideate step. It is where one takes the information and feedback from the empathy phase and identify a key problem and define it as an opportunity to brainstorm solutions to in the ideate phase. See *design thinking*.

Design Thinking – A design process model popular in the tech field. It is comprised of a cycle of five steps: empathize, define, ideate, prototype, evaluate/test. With product refinement achieved by rerunning the cycle. It reinforces the need of the user as the primary drive for design and ensures a strong link with the user through the design process. See also *empathize, define, ideate, prototype,* and *evaluating/testing*.

Direction – Sprites in Scratch have a direction property. This tracks or determines what angle they will move on if they **[Move (#) Steps]** and, depending on the rotation style property for the sprite, may orient their graphic to that degree angle as well. In Scratch, direction 0 is up, +90 is right, -90 is left. Sprites start with a default value of direction +90, or going right.

Duplicate – In Scratch, you can copy and paste code blocks, costumes, sprites, sounds, and backdrops. If you right-click on an object, you will get a pop-up of additional commands, generally allowing duplicating or deleting, but possibly others too. When you duplicate a code block, all code attached inside or under it will duplicate, too, and any typed or selected values will be copied as well. Duplicating a sprite will copy all its costumes, sounds, and code. There are many ways to copy things in Scratch. We give further explanation about it in Chapter 5, "Project 1: Dinosaur Dance Party".

Edge – In Scratch, the limit of the Stage Window view is called the edge. Objects collide with the edge the second any pixel of their costume touches it. The •[If On Edge, Bounce] code block will make an object that touches an edge bounce off it using the correct reflective angle. In many cases of basic movement, a sprite cannot move past the edge of the Stage Window. It will only move so far as some portion of its sprite is still visible in the Stage Window. This protection protocol can be worked around using Set X/Y movement, but •[Move (#) Steps] will be limited by it. The •[Go To [*random position*]] code block may move a sprite to colliding with the edge but will never place a sprite beyond the edge outside of the Stage Window. See also *Stage Window*.

Editor – The Scratch editor (or just editor for short) allows a user to create a new Scratch project or edit an existing one. It comprises of the Stage Window, stage, Properties panel, Sprite List, and then the left side of the screen depends on the tab selected, providing a workspace and tools for that tab. While the Scratch website also has the main page, search pages, studio pages, project pages, and others, we deal almost exclusively with the editor in this book because that's where all the code and digital art creation takes place. In general computing, an *editor* is any program that allows users to create or edit files, such as a word processing program, and is not limited to coding, but it can also apply to other software for coding.

Empathize – One of the steps in the design thinking product design process. *Empathize* is the phase where designers work with potential user groups to hear their experiences, problems, and needs. This gathers the information needed for the next step, *define*, where those problems are refined as opportunities for development. See also *design thinking*.

Evaluating/Testing – One of the steps in the design thinking product design process. *Evaluating/testing* is the final step, occurring after the prototype stage, where a product or service has been created and rolled out. Evaluating/testing determines how the product or service performs in the real world. Products can be further refined and developed by repeating the cycle, going to another empathizing phase to gather feedback from users after the evaluating/testing phase. See also *design thinking*.

Evaluation – In education, this is one of the six components of computational thinking. It is used to determine the effectiveness of a process or solution and to see if it can be generalized and applied to other problems. See also *computational thinking*.

Event – In Scratch, Events are a category of code block. Most events are hat blocks. Events are generally triggers that determine when code will be run. Without being attached to a trigger, code blocks will not be run. See *hat block* or *trigger*.

Execute – In computing, to *execute* means to run code. The computer is executing the commands you have given to it. This can also be referred to as run or running. In some programming languages, code must first be compiled (converted to direct instructions for the computer at the lowest level) before executing. Compiling is not necessary in Scratch.

Follow – Because online Scratch extends beyond just an editor, providing a platform for sharing content, user accounts can follow each other. When following another user, you get updates about their activity, such as when they share a new project. See also *Scratch account*.

Frame – In visual art (including video, animation, and video games), a *frame* is a single set complete drawing of the screen – all the data that comprises what is shown on the screen, or window, for a project. This is like a single picture in a film reel, or a single composited cell of animation. A frame represents the view of the project, animation, movie, or game world at that exact moment. A game, animation, or movie will have a frame rate that determines how rapidly a new frame is composed and drawn to the screen to create animation and update the user's view of the project.

Frame Rate – In visual arts, this is how quickly the graphical view of the project is updated, given in a number of frames per second (or FPS). Most computer imaging is done at 30 fps, or 30 frames drawn per second, though modern gaming has moved to 60 fps as a common standard. Cinema tends to run at 24 to 30 fps. The human eye and brain are believed to have a limit of comprehension and perception in the 30–60 fps range. This means we might not reliably notice things faster than that.

Function – In general computing, a *function* is a clearly defined set of instructions for the computer to achieve some particular purpose. It can refer to a built-in capability of a programming language, such as Scratch's [**Go To [*Random Position*]**], which automatically determines a randomized

X-coordinate in a range of -240 to 240, a randomized Y-coordinate in a range of -180 to 180, and then changes the sprite's X and Y properties to those new numbers. Functions can also be user-defined or custom, where the user creating a program can define a new function and provide all the instructions for it. This function can then be called in their code whenever needed, just like a prebuilt function can. The •My Block category in Scratch is the equivalent of building custom functions through the •[Define [My Block]] hat block and calling them through the custom stack block created for any •My Block that is created. Functions can have arguments or parameters that ensure needed numbers for the function to work are provided for it when it is called.

Grid – All computer graphical systems use graphs or grids to track the position and scale of objects they need to draw (show on screen). In Scratch, a coordinate system is used for the same purpose. It has its origin at the direct centre of the Stage Window, known as X:0 Y:0, and will centre a sprite. The Stage Window is 480 × 360 pixels, which gives visible coordinates from X: -240 Y: -180 to X: 240 Y: 180. X increases to the right, while Y increases to the top. Other platforms and digital systems may use different grid systems. One of the backdrops called "XY-grid" in the Backdrops Library can be used to see a diagram of the grid system in Scratch.

Hat Block – In Scratch, *hat blocks* are a shape of code block. They have a bump on the top that indicates nothing can attach above them. Hat blocks are all triggers, determining when the code attached below them should run. Most hat blocks are in the •Events category. Every script must start with a hat block to tell the computer when that code should be run. Without a hat block, the computer will never run the code. See also *event* or *trigger*.

Ideate – This is one of the steps in the design thinking design process. It follows the define phase, where the challenges or problems have been identified and defined as opportunities. *Ideate* is where possible solutions are conceptualized to meet the opportunities. After ideate, the prototype phase builds systems imagined by the ideate phase. See also *design thinking*.

Input – In general computing, input is feedback from the user or environment. It can be direct commands, such as key presses or mouse clicks; indirect commands, such as mouse movement; or passive input, such as sensor readings or clock ticks. Input is used to interact with the computer system, generally triggering some code or providing values or data.

Lag – In general computing, this slang is used to refer to delays in processes. Most commonly, it refers to any pause, disruption to Internet streaming for games, videos, or video conferencing. It can also refer to delays in stimulus–response systems or delays between decisions and results. Scratch can lag if you have too many clones running overly complicated code.

Language – In general computing, this can refer to either a user's human language or a programming language, a system developed to allow humans to give instructions to the computer that is more readable and convenient than giving more direct and literal machine instructions (also known as assembly). In Scratch, *language* refers to user language. In the editor, you can change the language Scratch uses by clicking on the "Settings" button in the upper left-hand corner of the screen. Scratch is available in 74 languages at the time of writing this.

Message – In Scratch, a *message* is part of the broadcast system. Messages are uniquely named user-defined triggers that can be used through three of the Event category code blocks. Messages are broadcast – a sprite broadcasts the message using the **[Broadcast [Message]]** code block, and then any and all sprites or stage with **[When I Receive [Message]]** code blocks will trigger and run their attached code. This is a very useful system to get multiple objects to respond to a specific event.

Modelling and Simulation – In education, modelling and simulation is one of the six components of computational thinking. It's about using models and simulations as both a method of understanding situations and processes as well as creating or changing situations and processes. Coding is an excellent way to utilize this method, as we can not only create models and simulations but also can run easily and efficiently even at massive scales, from which we can derive results, but we can also easily and affordably run enough simulations to derive statistical data for further analysis. See also *computational thinking*.

Nested – In general computing, nesting code is a method involved in certain functions where some code is run conditionally or in some way controlled by another function. It is contained by this controlling function and, in typed code, is generally indented from its controlling agent, which nests it inside, making it visually distinct in addition to its flow control aspect. In mathematics, we use parenthesis to nest certain math functions inside others to control the order of operations, similar to how nesting functions in coding. In Scratch, the C blocks nest code inside them, in some way controlling how the code inside them is run, such as conditionally with an **[If <*True*> Then]** code block or loop it like with a **[Forever]** code block.

Object – In general computing, objects are used in some programming languages as a model for how code is run and data is organized. An object is a discrete agent in the model, with properties that can be set or checked and have code run by it or affecting it. Infinite numbers of objects can exist in a model, and they can have different properties and behaviours. Scratch uses this system, with both sprites and the stage being an object. They each have properties, can have their own code, and can be affected by or affect others.

Parameter – In general computing, a *parameter* is a data point that helps define how a function should be run or is another form of data point that defines how the program, model, or simulation works. In Scratch, the Turbo Mode setting would be a parameter, as would the View Modes. The term is sometimes applied to any argument, value, or input that is passed to a function.

Pattern Finding – In education, this is one of the six components of computational thinking. It is also sometimes labelled as pattern recognition. It is the practice of analyzing data or systems to find patterns, either in behaviour or outcomes, or in similar processes and outputs. Pattern finding is often used to help standardize or simplify processes or solutions. See also *computational thinking*.

Pen Tool – In Scratch, the Pen tool is one of the optional extensions that can be added to Scratch 3, creating a new Pen category of code blocks. It provides the ability to draw lines or copy costumes to the background, on top of the current backdrop. The pen works by being set **[Pen Down]** (to draw) or **[Pen Up]** (to not draw) and then moved. The code is run by a sprite so that a sprite's movement will determine what is drawn if they have the **[Pen Down]**. If a sprite uses the **[Stamp]** Pen category code block, it will copy their current costume to the background.

Pixel – In general computing, a pixel is the smallest unit of graphical display, equivalent to a single point of colour. Computer monitors, TVs, and other digital displays are comprised of millions of pixels that display images by lighting up different colours. Digital art can be made in two different methods: pixel (defined by a grid of pixel colours) or vector (defined by mathematical instructions to construct shapes with given properties). All art is always displayed on the pixels of a digital display, though. In Scratch, the •**[Move (#) Steps]** block will move an object approximately (#) pixels on the screen, though this can be rounded off because of the angle of movement (see *direction*). The Stage Window in Scratch is 480 × 360 pixels, though in Presentation Mode, it is enlarged, so 1 pixel in-game will be multiple pixels on your display. See also *vector* and *grid*.

Player – In gaming, a *player* can mean either any top-level agent in the game capable of winning or scoring or, more specifically, a human playing the game. Players can be represented with avatars, or through characters, which are fictional agents within the game controlled by players. You can think of players as actors, and characters as the role the player takes in the game world. See also *character*.

Project – In Scratch, users create projects as the distinct applications or programs they can run and share. Each unique project is listed in the user's Scratch account. Each project gets its own project page, may or may not be shared, and can be edited in the editor and remixed or seen inside if shared. See also *project page* and *Scratch account*.

Project Page – Each project created in the online version of Scratch gets its own website page on the Scratch website. This page allows the creator to add instructions and notes to the project and to share it. If shared, the page can collect likes, favourites, and comments and allow others to play it, see inside the project, or remix it for their own use. A user's Scratch account links to all the project pages for all the projects they've created and saved.

Properties Panel – In the Scratch editor, the Properties panel is directly below the Stage Window and above the Sprite Listing. This panel allows you to see and edit the most commonly needed properties of a sprite. It has dialogs for showing/editing the name, X position, Y position, visibility, size, and direction of sprites, though sprites do have more properties than those listed.

Property – In computer science, a *property* is a piece of data assigned or associated with an object. It defines some aspect of its state in the simulation. Sprites in Scratch have many properties: name, X position, Y position, visibility, size, direction, costume, and more. Coding allows a creator to assess and alter properties of objects in their simulation, making them move, change colour, appear, or disappear. See also *object*.

Prototype – This is one of the five steps in the design thinking design process. It comes after the ideate phase, where the conceptual idea for a solution or product has been created. The prototype phase is the actual building of the product or service to bring the ideate phase's idea to life. It is followed by the evaluation/testing phase, where the prototype is tested in the real world to see how it performs. See also *design thinking*.

Remix – All projects in Scratch are under an open-source license as terms of use of the platform. This means that anything you create in Scratch can be shared with the world with no commercial use or restrictions on sharing. If you log in to a Scratch account, any project page you visit will have a Remix button on the top right corner. Remixing makes a copy of that project to your account that you can then edit and modify in any way you like, with the ability to share that project. This allows everyone on Scratch to learn from everyone else on Scratch since they can access the code they used to achieve their projects. We can all learn from everyone else through any shared project. See also *see inside*.

Reporter Block – In Scratch, reporter blocks are one of the shapes of code blocks. These are the round-edged pill-shaped blocks. They represent values, numbers, strings, or data. They allow a creator to refer to a data point somewhere in their project, such as **(X Position)** or **([direction]**

of [sprite1]). They can also be typed-in data, such as ●((#) x (#)). Any of the white oval spaces in any other code block can be filled either through typing in a value or through using a reporter block.

Reticle – In the Art tab, the canvas has its centre point marked by a reticle, or target symbol. It can be important to align your costumes in reference to the reticle, as this is the point which a sprite is based. The X and Y properties will align directly with the reticle position of the costume. As well when rotating, the rotation of the sprite will be centred on the reticle. When moving vector shapes (or groups) on the canvas, they will automatically snap to the reticle if they are brought close enough to assist with properly aligning things.

Scratch – Scratch is a coding platform built with an emphasis on primary education and interactive media. It was developed by the Media Lab at MIT and is currently maintained by the Scratch Foundation. Scratch uses block coding to allow easy, friendly access for even young students. It is not just a platform for creation but also for sharing through the use of user accounts, studios, project pages, and a searchable listing of all shared projects created in Scratch. It is currently in its third iteration, Scratch 3. You can try Scratch at http://scratch.mit.edu/.

Scratch Account – The Scratch website is not just a platform for creation but also a platform for sharing. It has an account system for users to organize all their projects, save them on the cloud, and share them with the world. Scratch accounts are not required to use Scratch but are a great benefit to users as they provide a cloud saving for projects, an autosave feature (when logged in), and the opportunity to share projects with others.

Scratcher – A creator that uses Scratch can be referred to as a *Scratcher*.

Screen Refresh – In general computing terms, *screen refresh* refers to clearing the draw buffer in an operating system and rebuilding the graphical data to display to the user. Typically, this is done at every 30th of a second or more on a computer. In Scratch, "Run Without Screen Refresh" is an option available when creating a ●My Block. If checked, the code runs differently than usual and will not wait one frame between repeats and other similar delays in normal code. This can allow you to create custom functions with ●My Blocks that can rapidly process data, set up levels, or other things. See also *frame, frame rate*.

Script – In general computing, a *script* is a separate sequence of code that can be called as needed from any other code in a project and is a way to compartmentalize the reused functions required for a program. In Scratch, a *script* is any stack of connected code blocks. A script must start with a hat block so that the computer knows when to run that code due to a triggering event.

See Inside – In Scratch, every shared project can be viewed inside and out by other users. On any project page, there's a See Inside button in the upper right-hand corner. If clicked on, it will open the Scratch editor with a copy of the project loaded. Users can see all the code, sprites, costumes, backgrounds, and sounds that were used to make the project. This allows users to learn how to make any of the projects they explore, turning the site's ability to share into the ability to have millions of users teach one another.

Shared – Projects in Scratch can either be shared or not shared by their creator. A shared project can be seen by other users on Scratch, who can find it in search results, visit its project page, and even see inside the project or remix it. An unshared project cannot be seen by anyone except the user that created it.

Size – In Scratch, all sprites have a size property that determines how large they display in the Stage Window. Size is a percentile, with 100 indicating normal size relative to its current costume's size, 50 indicating half size, and 200 indicating double size. Costumes can be checked for their base scale in the Costume tab's art workspace.

Sound Library – In Scratch, when the user has selected the Sound tab in the editor, the Sound Library can be accessed by clicking on the "Choose a Sound" button in the bottom left-hand corner. The Sound Library is a collection of music, notes, and sound effects available to users to use built-in to Scratch. In the Sound Library, sounds are listed as tiles with a name and a play button that if the user hovers their mouse over it will play the sound effect. In addition, at the top of the Sound Library, there is a search bar and categories that will only list associated sounds. Clicking on a sound will add it to the currently selected sprite (or stage) in the project.

Sound Listing – In Scratch, when the user has selected the Sound tab in the editor, the Sound Listing is shown on the very left-hand side of the editor. It lists all the sounds currently added to the selected sprite (or stage). The user can select any of the sounds by clicking on them and can then edit or listen to them in the workspace. The user can click, drag, and reorder them in the Sound Listing.

Sounds Tab – In the Scratch editor, this is the third of the three tabs. It allows users to add sound effects and music to the currently selected sprite or stage or to edit already-added sounds. The workspace in the Sounds tab allows to play the sound clips as well as apply a number of different transforms or effects to either the whole sound or to portions of it by clicking on the sound wave to indicate starts and stops. The Sounds tab is explained in detail in Chapter 4 in Book 1 – Beginners.

Sprite – In Scratch, sprites are the main workhorse of projects. They are the individual objects that appear in the project through their costumes, take action through their associated code, and make sound with their associated sounds. Each sprite is its own discrete agent operating code-independently but can interact and react to the other sprites. All the sprites in a project are listed in the Sprite Listing, while the currently selected sprite displays its properties in the Properties panel. Only one sprite, or the stage, can be selected at a time, and whatever is selected is the target for any work done in the workspace, whether code, art, or sound. In general computing, the term *sprite* refers to a piece of pixel art that will be displayed to represent an object, rather than being an object in its own right with swappable costumes.

Sprite Library – In Scratch, if the user decides to add a new sprite, they can click on the Choose a Sprite button in the bottom right-hand corner of the editor. This will show the Sprite Library, a collection of already-drawn objects available to all users of Scratch. In the Sprite Library, if a user hovers their mouse over one of the tiles, they may see it animate, showing the multiple costumes associated with that chosen sprite, but not all sprites have multiple costumes, so not all will animate.

Sprite Listing – In Scratch, all the sprites that have been added to a project appear in the Sprite Listing. This is found under the Properties panel, which is under the Stage Window in the bottom right-hand corner of the editor. Each sprite in the game is listed here in tiles, with a thumbnail displaying the current costume active for that sprite. The user can right-click on a tile to duplicate it or delete it. They can also click and drag sprites to reorder them in the Sprite Listing. The currently selected sprite will be outlined in blue, and a trash bin icon will appear in the upper right-hand corner of the tile that can be clicked to delete it. Whatever sprite is currently selected will display its properties in the Properties panel above, as well as show its associated code, costumes, or sounds in the workspace to the left.

Stack – In Scratch, a *stack* is a set of interconnected code blocks. The official name for a stack is "script", but *stack* is very commonly used. See also *script*.

Stack Block – In Scratch, the stack block is the most common shape of code block. These are the basic rectangular code blocks. They can fit above or below other code blocks and often have a value space that the user can type in a number or text into or can use a reporter block to fill to have it populate with a dynamic value from the project. Stack blocks connect into scripts, and each script must start with a hat block in order for the computer to know when to run the code.

Stage – In Scratch, the *stage* is a special object that handles the background in projects. It can have its own code, graphics (called backdrops), and sounds. The stage has limited code blocks that can be assigned to it because it cannot move, change layering or size, among other limitations. It is always directly centred in the Stage Window and is always the exact same size. The stage cannot be removed or deleted from a project.

Stage Window – In Scratch, a Project is displayed to users through the Stage Window. This appears in both the project page and in the editor. The Stage Window is a view into the simulation created in Scratch, whatever form it has taken – game, music video, interactive story, etc. The stage will always provide the background of the Stage Window since neither the view the Stage Window provides nor the stage is able to move. The Stage Window is 480 pixels wide and 360 pixels tall. Its border is called the edge. Whenever the project is played (through clicking the ▷), all the action and interaction will occur in the Stage Window. See also *edge, grid, project*.

STEAM – In education, STEAM is an acronym standing for science, technology, engineering, art, and math. It represents the earlier term STEM with the addition of *arts* to highlight the need for creativity. STEAM and STEM are terms associate with a push to highlight knowledge of the physical sciences and applied science in both formal education and after-school programs or hobbies. Coding is often used as a way to incorporate technology and/or engineering into STEAM curriculum.

STEM – In education, STEM is an acronym for science, technology, engineering, and mathematics. It is a common term used to highlight a focus on physical and applied science in education. Coding education has often been used as a major focus for STEM programs as a way to incorporate both technology and engineering. STEM is sometimes extended to STEAM with the addition of *art* as an additional focus.

Step – In Scratch, a *step* is a measure of distance, roughly equivalent to 1 pixel. In computing, it will often be used as a reference to the execution of the sequential order of code, with each line of code (or code block in Scratch) equating one step of execution.

String – In general computing, a *string* refers to a sequence of characters or typographic symbols. Words, sentences, and passwords are all strings. They are a form of variable or data point that is not numeric – therefore cannot have math operations performed on it. See also *value* or *variable*.

Touching – In Scratch, sprites can be tested if they are touching other sprites or colours, or if colours are touching other colours. *Touching* in this case is determined by the costumes of sprites involved, or the sprite involved and the colour of the background, or any other sprites as composited into a frame at that time. It is determined if any pixel, or pixels, of the assigned colours overlap in position. In general computing, *touching* is known as colliding. Importantly, touching only counts overlapping, not adjacency.

Trigger – In general terms, a *trigger* is anything that sets into motion an action or reaction. In coding, this is often an input (such as a key press) from the user or an input from a sensor system. The hat blocks in Scratch are examples of triggers; they are specific events that will cause Scratch to recognize their occurrence and can then be used to have the computer run associated code assigned under them. See also *hat block*, *event*, or *input*.

User – In general terms, a user is anyone using an application, program, or project. This can refer to either the person playing a project in Scratch (who can also be referred to as a player) or as the person using Scratch to create a project (who can also be referred to as a creator or Scratcher). See also *creator* or *player*.

Values – In general computing, a *value* is a data point. It can be assigned to a variable or required for a function or found in a data structure. The term is used widely and freely to refer to any kind of data point and may refer to an input or parameter. In Scratch, a value is a white oval in a code block, a data point required for the code block to do its job, helping define or quantify its actions. Values can be represented through reporter blocks, and reporter blocks can fit into value places in code blocks. See also *input*, *parameter*, *variable*, or *reporter block*.

Variable – In general computing, a *variable* is a memory assignment for the computer. It creates a reference name to a piece of data that the computer will hold in memory, which can be changed or referred to at will. By making a general reference, this data can be referred to at any point or can be modified as needed. In some computer languages, variables have set data types – integer or string, for example; in others, they are dynamic. In Scratch, •Variables are used the same as in general programming as fully dynamic data points, but it is also a category of code blocks that are used to work with •Variables. To work with a variable in Scratch, one must first go to the •Variables category and click on the Make a Variable button to create it first. See also *string*.

Vector – In general computing, *vector* art is one of two forms of art, the other being pixel (also known as raster). In pixel art, the canvas is defined as a grid of cells, with each cell representing a single pixel with an assigned colour. In vector art, the art is defined by mathematical formulae and instructions on how to build dynamic relationally positioned and proportioned shapes, lines, and spaces. Vector art, therefore, scales perfectly without any loss of quality and can have true curves and smoothness, unlike pixel art. It is, however, more difficult to create complex designs with. Scratch can create and edit both forms of art. It is highly recommended you try learning both and encourage students to do the same. We talk more about the two forms of digital art in Chapter 4 in Book 1 – Beginners. See also *pixel*.

View Mode – Scratch has multiple ways of presenting projects to users. By default, either on the editor or the project page, you will see the Stage Window at normal scale. In the editor, you can also choose between three view modes – default, compact (the Stage Window is half size, so you have more room to code), or presentation, also known as full screen. In Full Screen mode, the Stage Window will expand to fill as much room on your screen as possible. You can access the Default or Full Screen mode in either the editor or the project page. Full Screen mode can help students remember that they can't count on users being able to click on code blocks to execute them and must make full controls that don't rely on the user being able to manipulate code or sprites as an editor. See also *Stage Window*.

Workspace – In the Scratch editor, the workspace is the largest central part of the editor. Depending on which tab is selected, it is where the user can create their code, create or edit art for costumes or backdrops, or listen or edit sound. It will change nature with appropriate tools, depending on the tab selected. The content of the workspace will depend on the sprite or stage selected. See also *art workspace* or *coding workspace*.

X – In Scratch, X is both a position property of sprites and a dimension of the grid used to determine position and scale within the Stage Window. X measures the left–right positioning. X: 0 is the centre of the Stage Window, with X increasing to the right. The Stage Window is 480 pixels wide, making grid positions range from X: -240 to X: +240. See also *Y*, *grid*, or *Stage Window*.

Y – In Scratch, Y is both a position property of sprites and a dimension of the grid used to determine position and scale within the Stage Window. Y measures the up–down positioning. Y: 0 is the centre of the Stage Window, with Y increasing to the top. The Stage Window is 360 pixels tall, making grid positions range from Y: -180 to Y: +180. See also *X*, *grid*, or *Stage Window*.

Index

Green Text – entry is in Book 1
– Beginners
Blue Text – entry is in Book 2
– Intermediate
Red Text – entry is in Book 3
– Advanced

A

accessibility 16–17, 19–21, 7–8
accounts 21, 31, 153–154, 158, 193–195
add a background 69–70, 87–89, 24, 39–40
add a costume 90, 114, 58, 64
add a sound 61, 65, 71–72, 77, 80, 98, 131–133, 113–114, 64–65, 138–140, 146–150
add a sprite 47–48, 72, 89, 104, 106, 112, 23–24, 26–27, 45–48
ADD/ADHD *see* sensory issues
advanced students 148, 155, 151, 159–160, 184, 191, 196
AI 1, 11, 24, 76–78, 83–85, 108–109, 131–134, 142–143
algorithms: flow of code 146–147; grid marching 26–29; list stepping 17–19; place sorting 154–158
ammo/use limits 123–124, 138, 165–167
analogies, helpful 110–111, 60, 149–150
and logic 115–116, 52–55, 61–62
animation: cycle 72–74, 82–83, 34–35, 110–111, 128–131, 163–165; frame 62, 72–74, 78, 104–105, 121, 42, 52–53, 58–59; Fx 99–100, 58–61, 119–123, 146–150; motion 78–79, 92–94, 105–106, 108, 48–49, 58–61, 65–68, 138–140; reactions 117–121, 75–79; techniques 95–97, 41–42, 45–48, 34–35, 138–140, 163–165

animation projects 68, 85, 11, 32, 28–68
answer *see* ask
art: colour effects 29–32, 82–84; costumes (*see* costumes); digital skills 22–24, 146–148, 163–165; formats 50–51, 62; glow effect 112–124; hidden objects 58–61; invisible components 17–19, 75–79, 125–127; making in scratch 47–56, 87–91, 106–108, 112–113, 114, 23–24, 26–27, 74–75, 80, 88–89, 98, 100, 104, 106–107, 112–113, 117–119, 122, 124, 128, 129–130, 16–17, 23–26, 36–38, 58–61, 65–68, 71–72, 75–107, 110–117, 119–123, 125–140, 146–161; making patterns 20–21, 21–23; rotating 49; shape 87, 107, 11; symmetry 107, 23–26, 112–114, 119–123; vector 88–91, 107, 23–24, 88–89, 98, 104, 65–68, 69–107; working with backgrounds 87–89, 24, 125–127; working with sprites 47–56, 89–91, 107–108, 112–113, 24, 26–27, 56–57, 58–60, 62–63, 87–88, 98–100, 106–107, 118–119, 121–122, 16–17, 23–26, 30–31, 36–38, 58–61, 65–68, 71–72, 75–107, 110–117, 119–123, 125–140, 146–161; working with text 112–113, 80, 100, 121, 65–68, 104–105, 105–107, 146–150, 151–153, 154–158, 158–161
Artificial Intelligence *see* AI
art projects 85, 11, 69–107, 108–161
art tools 51–54; *see also specific tool names*
ask 35–39, 43–45
assessment 149–150, 154–155, 150, 152, 159, 180–182, 184, 196
assignments 31–33, 144–145, 148, 151, 184

attacks 112–114, 128–131, 135–137, 140–142
autism *see* sensory issues
automation 11, 20–23, 12–13, 32–33, 176–179
autosave 154, 158, 193–195

B

back *see* layers
backdrop: tab 47, 59–61, 87–89, 24, 58–61; blackout 24, 110; setting 69–70, 87, 24, 39–40, 96–97, 30–31, 36–38, 38–44, 48–56, 58–68
background music 71–72, 83–84
backgrounds, scrolling 125–127, 163–165
backpack 76–77, 157, 158, 162, 193–195
back up plans 164–166, 169–170, 205
backward *see* layers
behaviour, code 47, 61, 155–157, 148–150, 159–162, 196–198
best practices: commenting 149–150, 152, 185–186; development aids 36–38; encapsulating 116, 60; initializing 105–106, 111, 8–9, 15–16, 34, 39, 91–94, 113–115, 123–124, 36–38, 79–81, 85–88, 110–101; instruction 155, 160–161, 196–198; reset 94–95, 14–15, 14–16; troubleshooting 152, 155, 157, 193–207; visibility 97–98, 117, 124–132, 114–117
bias 12, 16, 6–7, 7–8
biology 102, 117, 21–23, 176–179
bitmap 50–51
block coding concept 17–20
bonuses/boosts 96–98, 123–124, 93–94
Boolean 43–44, 46; *see also* if
boss enemies 138–145
bounce, if on edge 120, 114–117
brightness 54, 89
broadcasts *see* messages
brush tool/size 51–52, 90–91
bugs/bug hunting 27–28, 97–98, 152, 157, 182–183, 193–207

buttons 55–57, 124–127, 129–132, 142–143, 65–68, 151–153, 158–161

C

canvas, art 48–52, 54–55, 87–91, 107–109, 24, 104–105, 17–19, 75–79, 125–127
careers 10–11, 6–7, 142, 7–13, 108–109, 162, 180–181, 208–210
categories, code block 41–42, 41–47
center *see* reticle
challenges, for advanced students 148, 151, 9–11, 184, 191
change X/Y *see* X/Y
character 104–105, 32, 78–80, 98, 28–68, 30–31, 69–107, 72–73
choice, player/user 15–16, 43–45, 73–74, 85–86, 129–132, 28–68, 158–161
circle tool 51, 53, 74–75, 88–90, 119–123
clamping 85–86, 135–136
clipping, sounds 64–65
clones: child/parent difference 125, 164–165, 76, 168, 135–137, 204–205; clearing out 91–94, 96–98, 143–144, 112–117, 128–131; for difficulty scaling 123–124, 140–142; procedural generation 106–109, 83–85, 97–101, 114–117, 123–125, 128–131, 131–134, 140–142; for projectiles 71–72, 112–114, 128–131, 135–137; techniques 96–98, 143–144, 135–137, 175–176; troubleshooting 164–165, 168, 204–205
cloud saves 31, 154–158
cloud variables 108–109, 154–158, 165–167
code blocks 38–39, 41–47, 107–109, 146, 148, 14–16
code disappeared 40–41
code flow *see* flow of code
code for the stage 71–72, 39–40, 48–50
code tab 38, 39–40, 60
coding is for everyone 1–2, 10–14, 169–170, 6–7, 148–150, 173–174, 7–8, 179, 208–210

coding problems 154, 159–160, 196
coin toss *see* randomization
collaboration 149–150, 154–155, 150, 152, 159–160, 180–186, 193–207, 208–210
collectables *see* power-ups
collision-based movement 109–111, 94–97, 167–172
collision correction 117–121, 73–75, 94–97
collision masks 75–79, 169–172
collisions: among clones 96–8; with changing objects 97–103; colours for terrain 111–112; for gravity 73–74; invisible or hidden 75–79; mutual destruction 117–119; projectiles 76–78, 117–119, 128–131, 135–137; for transitions 38–44
colour 19, 21, 41–42, 46–47, 54, 89, 91, 163, 29–32, 111–112, 112–113, 135–137, 201–204, 82–84, 201–204
colour, pen extension 29–31
colour collisions 163, 111–113, 117–121, 139–140, 168, 204–205
colour effects 63, 29–31, 82–84
colour panel/picker 47, 54
colour selection 163, 111–113, 168, 204–205
commenting 27–28, 149–150, 152–153, 9–11, 185–186
community 2, 4, 25–28
comparison, numeric (<=>) 126–127, 131, 51–52, 71–72, 85–88, 115–116, 123–124, 32–35, 48–50, 51–62, 75–88, 94–101, 104–105, 112–161
comparison, text 35–38, 43–45, 38–44, 45–50
compositing, layers 54–55, 56, 62, 88–89, 98, 100, 104, 106–107, 121, 112–117, 119–123, 138–140, 146–150; *see also* grouping
computational thinking 14, 22
computer opponents *see* AI
computer science 22, 24–25, 6–7, 140, 146, 148–150, 173–174, 1–16, 19–21, 165–167, 176–179, 182–183, 208–210

concurrence 86, 140, 162, 91–94, 201–204, 201–204
conditional statements *see* if; until
confused pairs 157–160, 162–165, 198–201
constant learning 5–6, 13–15, 169–170, 6–10, 146, 173–174, 3, 176–179, 179–180, 208–210
control blocks *see* if; repeat; until
controls: dragging 136, 15–16; keyboard 77, 80, 83, 92, 95, 105, 13–20, 23, 25, 27–30, 35–36, 85–86, 106–111, 123–124, 138–139, 14–17, 36–38, 72–73, 74–75, 94–97, 110–111, 112–114; mouse 113, 121–122, 124, 52, 54–57, 71–73, 76, 100–101, 125–126, 129–132, 30–31, 36–44, 50–56, 65–68, 105–107, 151–161; text input 35–39, 139; variables 18–19, 21–23, 72–75, 129–132
control structures, troubleshooting 161, 166–167, 201–204
conversations 129–131, 45–48, 52–55, 64–66, 50–58
cooldown 121–124, 112–114
coordinates *see* X/Y
copying code 75–77, 93, 157, 161–162, 196–198
costumes: about 48, 58, 61–63, 65; altering 62; animating 72–74; collision masks 169–172; creating 89–91, 106–108, 23–24, 26–27, 74–75, 85–88, 97–101, 98–100, 100–101, 104–107, 112–113, 116–119, 121, 124, 128–130, 16–17, 23–26, 36–38, 58–61, 65–68, 71–72, 75–107, 110–117, 119–123, 125–140, 146–161; mixing sprites 56, 58, 63–64, 30–31; movement-based 75–79; randomizing 27–29, 89; size limit workaround 104–106, 17–19; as sprite property 139, 143; as states 95, 117–121, 69–70, 163–165; tab 47–56, 89–91, 104–105, 23–24, 146–148

creativity 4–6, 12–13, 22–23, 25–26, 62, 169–170, 146–148, 173–174
critical thinking 13–15, 22, 169–170, 4, 148–150, 173–174, 9–11, 180–183, 208–210
CS *see* computer science
CSV files 12–13, 27
Ctrl + C/X/Z 56, 63, 75
curved (vector) 48, 53, 87–89, 74–75

D

damage systems 115–121, 117–119, 128–131, 131–134, 140–142, 144–145
data: about 45–47, 57, 110–111, 137, 12–16; processing 171–172, 12–27, 165–167; visualization 12–27
deaths 109–110, 69–70, 82–88, 119–123, 163–165
defined count 18–23, 27
delete 39, 45, 52, 57, 63, 72, 86, 88
depth *see* compositing, layers
design thinking 13–14, 175–176, 180–181
dialogue *see* conversations
difficulty scaling *see* game balance
digital citizenship/literacy 5, 11–12, 6–7, 208–210
direction 37, 63–64, 120, 137–139, 14–17, 19, 48–50, 71–75, 83–85, 106–111, 63–64, 75–79, 97–101, 114–117, 128–140
displays 110–116, 18–19, 21–22, 56, 74–75, 80–83, 121–122, 124–132, 142–143, 12–27, 36–38, 85–88, 151–161, 163–165
download 27–28, 29, 31, 154, 169, 206
drag mode *see* controls, dragging
duplicate code 75, 93, 16, 21, 56–57, 61–62, 125–127
duplicating sprites 100, 56, 83–85, 143–144, 23–26, 38–44, 51–61, 91–92, 131–134, 154–161

E

edge 37–38, 107–110, 120, 125, 17–18, 109–111, 73–74, 82–84, 112–117, 125–127, 131–137

encapsulating *see* best practices, encapsulating
end game 110–111, 114–116, 125–131, 48–50, 66–68, 98–100, 116–117, 61–65, 10405, 146–150
enemies *see* opponents
engagement 14–15, 21, 154–155, 142–143, 146–150, 159–160, 7–8, 175–176, 196
environmental effects 105–106, 94–96, 111–112, 73–74
erase 51, 53, 14–15, 29–31, 119–123
errors: as assessment tool 154–155, 160–162, 164, 8–10, 157, 159–160, 165–169, 179–182, 196, 201–205; growth mindset 154–155, 164–166, 8–10, 135–136, 146–150, 157, 159–160, 165–169, 173–174, 179, 181–182
evaluation, 13–14, 182–183; *see also* errors, as assessment tool; if; until
events blocks 42, 47, 71
events, broadcast messages *see* messages
events, concepts 139–141, 146–147, 134–135, 140–141
exceptions 114–116, 125–131, 161, 8–10, 17–18, 35–38, 43–45, 51–52, 81–83, 85–86, 96–98, 104–106, 109–112, 115–121, 123–124, 135–136, 138–141, 143–144, 148–150, 157, 9–11, 21–22, 32–35, 45–48, 50–62, 73–79, 94–97, 112–114, 119–123, 181–182, 193–207
exploration 30, 144–145, 148, 154–155, 8–10, 151, 159–160, 173–174, 28–68, 69–107, 176–179, 196
explosions 95–96, 128–131, 146–150
extensions 24, 148, 12–13, 151

F

fade transitions 57, 99–100, 41–42, 48–50, 58–61, 80–81, 64–65, 85–88
false 44; *see also* if; until
fill 51–52, 54, 87–89, 89–91, 106–108, 112–113, 23–26, 74–75, 88–89, 98, 104, 17–19, 75–79, 114–117, 119–123, 125–127, 128–131

flow of code 47, 139–141, 146–147, 160–161, 148–150, 201–204, 180–182, 201–204
following 26–28
font 53, 112–113, 78
forever 71, 72, 106, 108, 119–120, 125, 41–42, 53, 74–75, 83–86, 106–109, 36–44, 65–68, 72–79, 85–92, 97–103, 110–123, 125–127, 131–142
forever, pausing/stopping 114–116, 125–129, 52–55, 81–83, 115–116, 82–84
forward (layering) *see* layers
frame animation 48, 72–74, 92–93, 139, 140, 160–161, 117–121, 165–167, 34–35, 79–81, 85–88, 94–97, 125–131, 163–165, 201–204
frame rate 74, 92–93, 141–142, 113–115, 117–121, 123–124, 138, 73–74, 94–97, 112–114, 119–123
freezing 153, 158, 193–195
front (layering) *see* layers
functions 42, 60–61, 81–83, 91–94, 135, 7–11, 94–97, 167–169

G

game balance 118, 83–85, 87–91, 96–98, 104, 111–112, 117–121, 143, 153, 155, 71–72, 82–88, 101–103, 108–109, 112–114, 138–140, 158–161, 165–167, 175–176, 187–189
game over 109–110, 114–116, 125–130, 8–10, 48–50, 66–68, 98–100, 116–117, 138, 140–141, 2, 82–84, 114–117, 119–127, 140–142, 144–150, 187
game projects 102, 117, 144–145, 32, 69, 102, 28–68, 69–107, 108–161
game start 71, 105–106, 108, 111, 113–115, 119–120, 122–126, 34, 38–39, 41–43, 91–101, 105–108, 112–114, 121–127, 48–50, 63–68, 79–81, 105–107, 110–123, 125–127, 138–140, 151–153, 158–161
game states: concept 125–131, 161, 80–85, 115–124, 140–141, 165–167, 171–172, 4–6, 36–38, 50–62, 82–92, 119–123, 146–161, 180–181, 201–204; damage 117–121, 82–84; lives/death/game over 109–110, 125–129, 130–131, 48–50, 66–68, 98–100, 116–117, 61–62, 82–88, 91–92, 104–105, 119–123, 146–150, 172–175; narrative paths 35–36, 43–48, 51–52, 55–57, 28–29, 36–38, 51–62; play & pause/turns 80–85, 115–122, 140–141, 69–70, 82–88; waypoints & levels 89–92, 104–105
gate systems *see* key/lock systems
generating enemies 123–124, 76–78, 83–85, 87–88, 143–144, 158, 168, 108–109, 114–117, 123–125, 128–134, 138–140, 165–167, 175–176, 193–195, 204–205
generator objects 123–124, 71–72, 76, 83–85, 96–98, 143–144, 97–101, 114–117, 123–125, 128–142, 165–167, 175–176
ghost effect 63, 99–100, 41–43, 48–50, 58–61, 80–81, 64–65, 85–88, 101–103
Glide 45, 67, 158, 162, 53–54, 162–163, 167, 30–33, 55–58, 63–64, 85–88, 138–143, 146–150, 198–201, 201–204
Go To 105–106, 108, 122–124, 20–23, 25, 27–28, 55–57, 74–75, 78–80, 83–94, 96–98, 105–107, 19–23, 38–48, 56–64, 75–84, 89–90, 94–97, 97–101, 110–117, 119–123, 125–153
Go To Vs Glide To 158–159, 162–163, 198–201
Go To X Layer *see* compositing
gradient 52, 54, 87–91, 106–107, 112, 114, 74–75, 88–89, 98, 100, 75–79, 112–117, 119–123, 128–131, 135–137
graphical effects 41, 99–100, 139, 143, 163, 4–5, 29–30, 42, 48–50, 58–61, 66–68, 80–81, 100–101, 143, 167–168, 2, 82–88, 101–103, 119–123, 125–127, 144–145, 163–165, 198–201, 204–205
graphics *see* art
gravity 105–106, 71–72, 4–6, 73–75, 167–172
green flag 37, 47, 71–72, 74, 112–113, 119–120, 123–124, 141, 15, 34, 45–47,

72–76, 85–94, 96–101, 106–107, 117, 121–122, 124–128, 168, 16–17, 23–26, 30–31, 36–44, 50, 52–55, 65–68, 71–72, 75–92, 94–97, 101–103, 105–107, 114–117, 128–131, 151–158, 204–205

green flag, restarting without 105–106, 112–113, 105–107, 112–117, 124–132, 110–111, 114–127, 146–153

grid *see* algorithms; grid marching; X/Y grouping 54–55, 62

H

handles (vector) 48, 53, 104
hat blocks *see* events; messages
hazards 106–109, 123–128, 83–85, 98–100, 117–121, 82–88, 128–131, 135–137
health 109–110, 125–129, 115–121, 82–88, 114–123, 172–175
hide *see* visibility
high scores 110–111, 122–123, 131, 87–88, 96–98, 113–117, 117–119, 128–134, 144–145, 154–158
hit points *see* health

I

If On Edge, Bounce 119–120, 125, 114–117
If Statements: game state 125–131, 81–85, 115–116, 50–62, 79–105, 119–125, 138–140; keyboard controls (*see* controls; keyboard); mouse controls (*see* controls; mouse); multiple choice 43–45, 32–33, 38–44, 48–52, 55–62, 154–158; multiple conditions 43–45, 115–116, 52–55, 61–62, 73–74, 117–119; properties 71–72, 32–35, 38–44, 48–50, 52–55, 82–88, 112–119, 125–131, 144–145, 151–153; random chance 26–29, 135–137; sprite collisions 109–110, 76–78, 81–94, 98–100, 107–111, 38–48, 73–75, 82–88, 91–101, 117–123, 140–142; text comparison 35–38, 45–50, 52–55, 82–84, 97–101; variable assessment 125–131, 81–86,

115–124, 129–132, 21–22, 32–35, 45–48, 50–55, 58–62, 75–90, 94–101, 104–105, 112–114, 119–125, 138–150, 154–161

If Vs If/Else 43, 146–147, 158, 160, 8–10, 43–45, 165, 198–201
inertia & friction 73–74, 94–96, 111–112, 69–70, 72–73, 75–79, 167–169
inheritance *see* parent/child concept
initialize *see* game start
input *see* controls; keyboard & controls; mouse
interface *see* controls; displays; variable displays
internet problems 20–21, 153, 164–165, 158, 169–170, 7–8, 193–195, 205–206
inventory 45–57, 28–29, 36–38, 52–61, 163–167, 172–175
invisibility *see* visibility
Item Use 45–48, 52–66, 123–124, 36–38, 52–55, 64–65, 165–167, 172–175
iteration 140–141, 146–148, 8–10, 20–23, 135, 138, 94–97, 167–169, 175–176, 184

J

jobs *see* careers
join text 129–130, 38–39, 41–43
jumping 105–106, 69–70, 74–75, 93–94, 167–169

K

keyboard input *see* controls, keyboard; key press event
key/lock systems 45–48, 52–57, 140–141, 28–29, 36–38, 50–62, 172–175
key press, testing 85–86, 106–111, 138–139, 36–38, 72–75, 91–92, 94–97, 110–114
key press event 47, 77, 82–83, 94–95, 105, 141, 143–144, 14–17, 19–23, 25, 27–30, 123–124, 138–140, 14–19, 36–38

L

language 17, 18, 22, 26, 34, 149, 36–38, 143, 148–150, 152, 171–172, 1–3, 7–8,

12–16, 22–23, 182–183, 185–186, 208–210
layers 41, 48, 54–55, 63, 113, 163, 41–42, 48–50, 56, 60–61, 76–77, 98–101, 104–109, 121–122, 124–129, 142, 167–168, 30–31, 52–55, 65–68, 71–72, 75–81, 85–88, 91–92, 104–107, 114–117, 119–123, 125–127, 140–142, 144–145, 151–153, 163–165, 169–172, 181–182, 204–205; *see also* compositing
level-based clones 96–98, 143–144, 97–101, 172–176
level-based costumes 104–106, 129–132, 134–135, 140–141, 71–72, 82–84, 91–94, 101–103, 172–175
level-based properties 79–81, 89–92, 94–97, 172–175
level systems 104–106, 116–117, 129–132, 134–135, 140–144, 171–172, 4–6, 69–72, 79–81, 91–107, 162, 172–176, 181–182
limitations: ammo/power up uses 123–124, 165–167; applying variable 85–86, 48–50; canvas and art size 35, 37–38, 89–91, 106–109, 146, 17–18; clone count 153, 158, 193–195; positions 17–18, 48–50; size 104–106, 135–136, 17–19; touching options 96–8; visibility & collisions 75–9; *vs.* text code 7–8
Line tool 52, 23–26
list variables 12–22, 27, 165–167, 172–175
lives *see* health; variables, health/lives
logic *see* computer science; game states; operators
logic, teaching 13–14, 17, 19, 22, 42, 117, 147, 161–162, 169–170, 8–10, 140, 142, 148–150, 165–166, 173–174, 9–11, 154–158, 181–183, 201–204, 208–210
Login 21, 31, 153–154, 158, 193–195
lookalike code blocks 157–160, 162–165, 198–201

looks blocks, troubleshooting 159, 163, 164, 167–168, 198–201, 204–205
loops 42–43, 71–72, 78–79, 108–109, 130–131, 140, 36, 71–72, 112–113, 134–135, 32–33, 94–97; *see also* forever; music; repeat; until
loss *see* game over

M

Make a Variable 110–111, 122–123, 125–126, 18–19, 21–23, 14–16
masks 69–70, 75–79, 169–172
mathematics: about 22, 28, 42, 44, 46, 50, 110–111, 19, 133, 136–138, 19–21, 35–36, 175–176, 176–179; formulae 18–19, 21–26, 72–75, 83–91, 94–96, 106–111, 117–121, 17–26, 35–36, 38–44, 55–58, 73–79, 94–101, 110–111, 140–142; geometry 11–31, 71–75, 17–23, 35–36, 74–75, 97–101, 110–111, 114–117, 142–143; multiplication 79, 82–83, 92–93, 99–100, 105, 108, 21–29, 19–27, 74–75, 93–94
maximums 18–19, 85–86, 21–22, 48–50
menus 112–116, 8–10, 124–132, 134–138, 140–143, 150, 9–11, 65–68, 105–107, 151–153, 154–158, 158–161, 181–182
messages: about 46, 47, 149, 8–10, 35–36, 13–15, 140–141, 150, 152, 185–186; cutscenes 45–48, 52–55, 58–66, 117–123, 51–55, 89–90, 140–142; game over (*see* game over); generation 119–123, 128–131, 138–142; interfaces 58–66, 76–77, 81–83, 124–132, 142–143, 21–26, 151–161; sequential questions 35–36, 43–45; start game (*see* game start); transitions (*see* transitions); troubleshooting 159, 163, 159–160, 165–169, 196–198, 201–204
minimums 18–19, 21–23, 85–88, 135–138, 17–19, 21–22
missed shots, deleting 71–72, 76–78, 87–88, 112–114, 119–123, 128–131, 135–137

motion blocks: arc motion 71–75, 74–75, 167–169; centric movement 119–120, 125, 11–26, 71–72, 106–111, 117–121, 85–88, 167–169; dynamic speed 73–74, 94–96, 109–112, 117–121, 123–124; gravity 105–106, 71–72, 73–74; random movement 95–97, 120–125, 23–26, 36–37, 87–91, 96–98, 97–101; rotation styles 34, 104–109, 30–31; stage 48–49, 146, 155–156, 160–161, 82–84; using variable input 18–23, 72–73, 83–85, 17–21, 30–33, 38–44, 55–58, 72–75, 89–92, 167–169; X/Y movement 79–82, 92–94, 26–29, 78–80, 87–91, 17–21, 30–33, 55–58 72–75, 82–84, 89–92, 97–101, 167–169
mouse controls 113, 121–124, 126, 144, 15–16, 52–66, 71–72, 76–77, 100–101, 124–129, 30–31, 38–44, 169–172
mouse over 47, 70–71, 90, 106, 112, 136–137, 23–29, 45–48
movement: cycle lockout 109–110, 114–116, 126–127, 130–131, 76–78, 83–85, 87–88, 45–48, 63–64, 73–74, 85–88, 94–97, 119–123, 128–131; cycles 105–106, 119–120, 125, 71–72, 83–85, 117–121, 30–33, 72–75, 94–97, 140–142, 97–103, 142–143, 167–169; flight 105–106, 121, 71–75, 110–111; goal testing 32–33; inertia & friction 73–74, 94–96, 111–112, 72–73; limits or stopping 109–110, 119–120, 17–18, 71–72, 85–86, 109–111, 117–121, 48–50, 73–74, 128–131; map 106–111, 38–48; patterns 79–82, 92–94, 146–147, 11–31, 48–50, 52–55, 63–66, 117–121, 114–117, 142–143; reversing 109–111, 73–74, 169–172; states 126–129, 109–112, 117–121, 32–33, 75–79, 82–88, 93–103, 142–143
multiple choice 43–45, 51–62
multiple condition handling 115–116, 61–62, 101–103

music: about 12–13, 24, 56–57, 61, 65, 70–71, 77, 80–81, 83–84, 136, 2, 12–13, 64–65, 146–153; looping 71–72, 64–65; projects 71–72, 28–68
mute 83–84
my blocks 42, 47, 58–61, 81–83, 91–94, 134–135, 9–11, 34–35, 73–75, 94–101, 128–131

N

narrative games 32–68, 28–68; see also paths
nested ifs 8–10, 32, 43–45, 109–111, 115–116, 123–124, 32–33, 38–52, 58–61, 74–75, 82–84, 91–97, 112–114, 119–123, 142–145, 154–158
nested loops 82–83, 139–141, 11, 20–23, 26–29, 144–145
nesting 126–129, 139–141, 43–45, 80–81, 83–85, 21–22, 142–143, 154–158
not logic 43–45, 109–112, 148–150, 36–38

O

object property referencing see referencing sprite properties
obstacles 88–91, 93–94; see also hazards
offline scratch 4, 20–21, 27, 29, 164–165, 158, 169–170, 193–195, 206
off-screen deletion 71–72, 112–114
off-screen protection 37–38, 17–18, 104–106, 135–136, 17–19, 125–127
operators 42, 96–97, 106–108, 120–124, 127–131, 158, 160–161, 4, 18–19, 23, 26–29, 36–39, 43–45, 71–75, 81–91, 94–98, 106–111, 113–121, 123–124, 129–132, 139–140, 148–150, 165–166, 14–23, 32–62, 73–105, 110–158, 198–204; see also and logic; not logic; or logic
opponents 123–124, 76–78, 83–85, 114–117, 123–125, 128–145, 175–176
or logic 161, 36–38, 138–140, 166, 73–75, 91–92, 94–97, 117–119, 131–137, 201–204

order of operations 38, 47, 139–141, 146–147, 160–161, 36–37, 91–94, 112–113, 134–135, 140, 148–150, 166, 21–22, 32–35, 38–48, 50–62, 73–79, 82–84, 94–101, 117–123, 135–137, 142–143, 154–158, 201–204
outline (art) 51, 54, 87–89, 48–52, 74–75, 88–91, 104–106, 112–113, 17–19, 75–79, 114–117, 119–123, 125–131
over-size 107, 104–106, 109–111, 17–19, 75–79, 125–127

P

paint, custom sprites 51, 62, 89–91, 106–108, 112–114, 137, 24, 26, 41–43, 48–50, 74–75, 80, 88–91, 98, 100, 104–107, 116–119, 121, 124, 128, 16–17, 36–38, 58–61, 71–72, 79–90, 93–107, 110–117, 119–123, 125–140, 146–161
paintbrush *see* brush tool
paint bucket *see* fill
parameter 7–8, 94–97
parent/child concept 123–124, 163–164, 71–72, 74–77, 143–144, 168, 16–17, 97–101, 108–109, 128–131, 135–137, 165–167, 172–176, 204–205
paths, branching 39–40, 43–45
paths, merging/converging 51–53
pen tool 11–31
perspective 106–109, 35–36, 71–72, 163–165
physics *see* gravity; inertia & friction
pixel size 35, 107–108, 26–27, 32–33, 125–127, 163–165
plagiarism 28, 32, 62, 8–10
platform (computing) 17, 24–26, 153–154, 135–136, 146–148, 169, 171–172, 1–3, 7–13, 206, 208–210
platforms, terrain types 93–94, 97–103
play sound *vs.* start sound 159–160, 164, 198–201
playtesting 37, 169–170, 48–50, 112–117, 181–182
points (score) *see* scores

points (vector art) 48, 50, 53, 87–89, 74–75, 98–100, 104–106, 128–131
popups, making 28–29, 36–38, 52–61, 163–165, 169–172
positioning *see* glide; Go To; X/Y
position limits *see* off-screen deletion; workarounds, offscreen movement
position testing 109–110, 119–120, 125, 71–72, 32–33, 73–74, 82–84, 112–117, 125–127, 142–143
power ups 96–98, 123–124, 93–94
predicting mistakes 154–164, 36–37, 159–169, 179–180, 196, 198–205
procedural generation 13, 108–109, 21–29, 87–91, 96–98, 108–109, 114–117, 123–125, 135–137, 162–165, 175–176
progression systems 35–36, 80–83, 112–113, 17–19, 28–68, 38–44, 50, 61–62, 89–92, 123–125, 138–140, 162, 172–175
projectiles 71–72, 83–85, 112–114, 119–123, 128–131, 135–137
project page 28, 32, 94–95
projects 8, 13–14, 23, 25–28, 35, 57–58, 144–145, 150–151, 165, 167–168, 1–5, 8–10, 133–134, 140, 144–156, 158, 169, 171–172, 1–11, 162, 176–186, 192–195, 206
properties 54, 58–59, 63–64, 79, 80–81, 97–98, 137–139, 156, 163, 8–10, 13–14, 15–16, 103, 105–109, 135–136, 138, 144, 160–164, 4–8, 108–109, 135–137, 165–167, 175–176, 182–183, 196–198, 198–201
pseudocode 165–166, 170, 9–11, 206–207

R

radial gradient 54, 91, 88–91, 114–117, 119–123, 128–131
random generation *see* procedural generation
random intervals 87–91
randomization: chance events 27–29, 123–125; data/variables 23, 14–16;

direction/movement 119–120, 123–124, 36–37, 83–85, 142–143; play dynamics 158–161, 175–176; position 108–109, 121–124, 25, 87–91, 96–98, 131–134

rate of fire (ROF) *see* cooldown

reactions 52–57, 83–85, 50–61

reading code 146–147, 149–150, 160–161, 148–150, 152, 159–160, 182–184, 196

readouts *see* variable display

rectangle tool 52–53, 87–88, 107, 112, 23–29, 43–45, 48–50, 100, 104–107, 116–117, 124–127, 16–19, 71–72, 75–79, 85–88, 97–103, 112–114, 125–131, 135–137

recursion 20–21, 36–37, 43–45

referencing sprite properties 106–109, 138, 22–23, 55–58, 97–101, 114–117, 128–131, 135–137, 165–169, 182–183

relative positions *see* referencing sprite properties

remixing 26–28, 32, 148, 153, 8–10, 151, 158, 1–3, 69–70, 193–195; *see also* inside

repeat loops: about 82–83, 92–93, 131, 140, 146–147, 94–97; animation fx 78, 96–97, 99–100, 41–42, 48–50, 80–81, 121–123, 85–88, 119–123, 144–150; animation movement 78–81, 92–94, 104–108, 15–29, 36–37, 48–50, 58–61, 71–72, 94–101, 125–127; cloning 153, 96–98, 158, 17–19, 123–125, 193–195; data processing 26–29, 17–19, 21–22; nested 82–83, 20–23, 26–29; pattern making 15–29; randomizing 26–29, 88–91, 14–16, 123–125, 146–150; timing mechanism 80–81, 92–93, 106–108, 17, 48–50, 97–101, 125–131; until clause 71–72, 112–113, 134–135, 32–33

replay 94–95, 105–106, 109–110, 112–116, 91–94, 104–106, 116–117, 124–127, 69–70, 85–90, 110–111, 146–153

reset 94–100, 123, 14–15, 29–31, 34–35, 48–50, 106–107, 113–115, 105–107, 110–111, 146–153

reshape tool 51, 53, 87–89, 139, 74–75, 98, 104, 117–119, 129–132, 146–148, 128–131

resize 48–50, 52, 87–89, 112–113, 150, 26–29, 62, 78, 104–105, 111–112, 152, 14–16, 35–36, 52–55, 128–131, 185–186

restart *see* replay; reset

reticle 49, 107, 74–78, 16–17, 23–26, 75–79

reverse movement 109–111, 117–121, 73–74, 94–97

rgb *see* colour

room-specific sprites *see* scene-specific actions

rooms *see* scenes

rotation 49–52, 63, 96–97, 142–143, 24, 34–35, 104–107, 30–31, 72–73, 75–79; *see also* direction

run once protection 45–48, 45–48, 50–61, 82–84, 89–92, 138–140

S

saving data/progress 38–39, 112–115, 129–132, 140–141, 89–92, 138–140, 172–175

saving projects 26–27, 31, 34, 57–58, 154, 164–165, 158, 169, 193–195, 206

say/think blocks 129–131, 159, 163, 32–68, 164, 168, 48–61, 94–97, 198–201, 204–205

scaling difficulty 123–124, 158–161

scaling, perspective 28–29, 35–36, 163–165

scene changing *see* transitions

scenes 35–36, 39–45, 104–106, 129–132, 38–48, 172–175; *see also* transitions

scene-specific actions 45–48, 51–57, 81–83, 115–123

scores *see* variables, scores

screen refresh *see* frame rate, my blocks

scripts *see* stacks
see inside 28, 94–95, 144
selecting a backdrop 59–60, 70, 87, 136–137, 39–43, 45, 96–98, 140–141, 148–150, 28–31, 36–44, 48–50, 65–68
select tool 51–52, 88, 91, 107, 48–50, 62–63, 80–81, 98–100, 117–121, 64–65, 82–84, 119–123
sensing blocks 42, 109–110, 15–16, 35–36, 76–78, 85–91, 106–109, 111–112
sensory issues 21, 164–165, 169–170, 206
sequences: animations 104–105, 121, 41–43, 48–55, 58–68, 121–123, 34–35, 65–68, 85–88, 110–111, 119–123, 128–131, 146–150, 163–165; code block connections 38–39, 41–47, 139–140; drawing routine 11–31
sfx *see* sound effects; special effects
sharing 25–29, 31–32, 35, 154
show *see* visibility
simulation 12–14, 22, 69–101, 102–132, 169–172, 176–179
size: animating 95–97, 48–50, 146–150; art 35, 48–49, 87–91, 107, 112, 26, 16–17, 125–127; limits, workaround 107, 104–106, 135–136, 16–17, 125–127; property 63, 104, 135, 13, 71, 76, 78, 104–106, 135–136, 72–73
social learning 13, 154–155, 169–170, 150, 159–160, 9–11, 193–207
sound, volume 83–84, 125–127, 138–140, 146–150
sound editing 56, 64–65
sound effects 77, 80–81, 98–99, 131–133, 58–61, 112–114, 64–65, 117–123, 128–131, 138–140, 144–150; *see also* music
sound issues 21, 159–160, 164, 125–127, 198–201
sounds tab 56, 61, 65, 71–72, 77, 98–99, 131–133, 135–136, 64–65, 117–123
special effects 99–100, 29–31, 41–43, 58–66, 78–81, 119–123, 128–131, 138–140, 144–150; *see also* graphical effects

sprite library 58, 61–62, 65, 72, 89–90, 104, 119, 123, 34–35, 39–40, 45–53, 55–57, 58–61, 63–66, 71–72, 76–80, 96–98, 124–127, 129–132, 30–31, 38–44, 50–61, 65–68, 72–73, 75–79, 114–117, 151–161
sprites 38, 40–41, 58, 60–65, 72, 75, 89–91, 104, 106, 112, 114, 119, 123, 136–137, 155–157, 13–16, 23–26, 26–29, 35–36, 58–63, 104–106, 135–136, 138, 143–144, 160–162, 167–169, 22–23, 75–79, 114–117, 135–137, 165–167, 175–176, 196–198, 204–205
stacks 42, 146, 150, 162, 60, 152, 165–167, 14–16, 110–111, 185–186, 201–204
stage: backdrops 58–60, 69–72, 87–89, 104, 119, 13–14, 23–26, 39–40, 104–106, 38–44, 82–84; code for 60–61, 70–72, 83–84, 104–109, 140–141, 14–23, 30–31, 48–50, 114–117, 123–125, 138–140, 158–161; as object 58–61, 70–72, 156, 24, 39–40, 140–141, 14–16; window 35, 36–37, 58–60, 70, 108, 120, 138, 13–14, 17–18, 23–26, 35–36, 87–88, 104–106, 111–112, 135–136, 30–31, 73–74, 82–84
start game *see* game start
start screen 112–114, 100–101, 142–143, 65–68, 105–107, 151–153
state machines: about 128–131, 141, 161–162, 115–116, 140–141, 171–172; character states 75–79, 82–84, 85–88; gameplay/pause 115–116, 117–121; levels 79–81, 91–92; lives & death 125–128, 82–88; menu/inventory 36–38, 151–161; player turns 81–83; rooms/scenes 38–48; story key/lock 45–48, 52–57, 28–29, 36–38, 50–62; timer locks 113–115, 112–114, 138–140; *see also* game states
step (movement) 38, 119–120, 125, 138, 13–23, 26–29, 48–50, 52–55, 71–74, 94–96, 109–111, 117–121, 112–117, 128–137

stop 37, 42, 44, 74, 109–110, 114–116, 125–129, 131, 48–50, 52–55, 63–68, 98–100, 63–64, 82–88, 104–105, 114–117, 119–123, 125–127, 138–142, 144–150
studios 26, 27
switches *see* state machines; key/lock systems

T

tabs *see* backdrop tab; code tab; costumes tab; sounds tab
terrain 111–112, 71–72, 93–97, 97–103, 169–172
terrain-dependant actions 74–75, 93–97, 167–169
testing text values 35–38, 43–45
text: coding 17, 148–150, 171–172, 1–3, 7–8, 14–16, 179, 208–210; as data 41, 43, 45, 36–38, 48–50; dialogue 129–131, 159, 35–36, 41–43, 45–55, 58–68, 87–88, 98–100, 164, 50–61, 63–64, 146–150, 198–201; input 35–38, 139; text, joining 129–130, 38–39, 41–43; text, shadow/highlight effect 112–113, 80–81, 98–101, 65–68; tool (graphics) 51, 53, 112–114, 48–50, 80–81, 98–101, 116–117, 121–123, 129–132, 23–26, 36–38, 105–107, 125–127, 146–150, 154–158
time lines 34–35, 97–101, 163–165
timer, limit 138, 97–101, 112–114, 128–131, 165–167
timer, stopwatch 112–117
timers 113–117, 138, 34–35, 97–103, 112–114, 128–131, 163–167, 175–176
timing, animation 72–73, 78, 80–83, 92–93, 107–109, 141–143, 158–160, 17, 39–43, 45–57, 80–81, 121–123, 164–167, 28–29, 32–35, 72–75, 94–103, 119–123, 146–150, 163–165, 198–204
title screen *see* start screen; menus
tools 51–54; *see also specific tool names*
touching *see* collisions

touching, limitations 162, 96–98, 167–168, 204–205
transitions: backdrop switch 39–40, 38–44, 61–62; boss battle 138–140; death restart 85–88; fade to black 41–43; level 79–81, 91–92; overlays 112–116, 81–83, 98–101, 116–117, 124–127, 36–44, 63–64, 104–107, 119–123, 125–127, 146–158; switching sides 45–48; title screen 112–114, 100–101, 124–127, 63–64, 105–107, 151–153
transparency, art 53–54, 89, 91, 112, 114, 26–29, 74–75, 104–106, 75–79, 112–117, 119–123, 125–127, 135–137
transparency, ghost fx 63, 99–100, 139, 163, 41–43, 48–50, 58–61, 66–68, 80–81, 168, 85–88, 101–103, 119–123, 144–145, 204–205
troubleshooting 146–147, 152, 157, 9–11, 179–180, 193–207
true *see also* if; until
turns 80–83

U

undo 34, 48, 56, 139, 109–111, 117–121, 74–75, 169–172
until 44, 46, 71–72, 126–127, 132–133, 158–160, 162, 71–72, 112–115, 134–135, 162–167, 32–33, 64–65, 138–140, 146–153, 198–204
uploading 22, 31, 34, 56, 61–62, 106–107, 125–127
user interface *see* controls; displays; variable displays

V

value (coding concept) 17, 44–47, 54, 139, 158–159, 161, 18–19, 23, 43–45, 72–73, 85–86, 106–109, 135–138, 163–165, 12–16, 21–26, 32–33, 74–75, 94–97, 181–183, 198–204
variable displays: for data 14–16, 23–26, 27; show & hide 98–101,

124–127, 129–132, 151–158; sliders 18–19, 21–23, 72–74; techniques 138, 142–143, 163–167, 172–176
variables: about 42, 46, 110–111, 122–123, 125–128, 136–138, 140–143, 19–21, 165–167, 172–176; health/lives 125–128, 82–88, 117–123; hiding/showing 38–39, 80–81, 98–101, 116–118, 124–127, 36–38, 151–158, 163–165; limiting 72–73, 86–87, 123–124, 138; player controls 18–19, 21–23, 72–74, 86–87, 106–109, 115–116, 72–74, 82–84, 167–169; randomizing 23, 26–29, 138, 14–16; scores 110–111, 122–123, 87–88, 113–115, 117–119; special types 12–13, 108–109, 154–158, 165–167; testing 122–123, 125–128, 45–48, 51–53, 112–113, 123–124, 21–22, 32–35, 50–62, 79–105, 123–125, 138–140, 142–143, 154–158, 172–175; and text 129–131, 32–38, 45–48, 56–58
vector art, using 50–51, 53, 54–55, 62, 87–91, 107–108, 112–115, 23–29, 48–50, 55–57, 62–63, 74–75, 80–81, 88–91, 98–101, 104–107, 112–113, 117–132, 16–19, 36–38, 65–68, 71–72, 75–107, 110–117, 119–123, 125–140, 146–161
victory conditions 87–88, 112–113, 116–118, 61–62, 104–105, 144–150
views 35, 40–41, 48–49, 90, 146, 104–106, 94–97, 167–169
visibility 63, 97–100, 110–111, 123–124, 139, 163, 23–29, 34–35, 39–40, 45–50, 52–57, 66–68, 74–77, 80–81, 88–91, 96–101, 112–118, 121–129, 135–136, 143–144, 167–168, 14–19, 36–44, 48–61, 63–68, 75–88, 93–107, 110–123, 128–131, 135–140, 146–161, 204–205
volume 41, 57, 83–84

W

wait 72–76, 80–81, 100–101, 141–142, 8–10, 21–23, 45–48, 76–78, 80–83, 87–88, 106–107, 112–129, 134–135, 138, 21–22, 34–35, 38–44, 55–56, 65–68, 79–88, 94–103, 105–107, 114–131, 135–137, 140–142, 144–145, 151–153, 158–161
walking 125, 36–37, 48–50, 30–31, 72–73
waypoints 112–113, 89–90, 172–175
win screen 66–68, 116–118, 63–64, 104–105, 146–150
workarounds: answers 38–39; clone-specific inheritance 135–137; drawing circles 17–18; maximum size 104–106; minimum size 17–19; mutual destruction 119–123; offscreen movement 125–127; run once collisions 45–48; touching options 96–98
workspace 36, 38–41, 48, 56–57, 64, 107, 139, 74–75

X

X/Y: changing 79–81, 92–93, 105–106, 108, 26–29, 58–61, 71–72, 78–80, 94–96, 32–33, 72–75, 79–81, 85–88, 94–101, 110–111, 114–117, 125–127, 131–134, 138–140, 142–143, 146–153; coordinate system 37–38, 41, 45, 49, 79, 112, 137–138, 41–43; glide to 36–37, 45–48, 52–55, 30–31, 55–61, 63–64, 85–88, 138–143, 146–150; position 49, 94–95, 71–72, 88–91, 19–21, 32–33, 35–36, 38–44, 55–58, 79–90, 112–117, 125–127, 128–137, 142–143, 163–165; setting 94–95, 26–29, 52–55, 114–117, 131–104; troubleshooting 156, 158–159, 161, 161–165, 196–204

Z

Zoom 40, 48, 90, 146

For Product Safety Concerns and Information please contact our EU representative GPSR@taylorandfrancis.com
Taylor & Francis Verlag GmbH, Kaufingerstraße 24, 80331 München, Germany

www.ingramcontent.com/pod-product-compliance
Lightning Source LLC
Chambersburg PA
CBHW060510300426
44112CB00017B/2611